Can precision medicine be personal;
Can personalized medicine be precise?

Can precision medicine be personal; Can personalized medicine be precise?

Edited by

YECHIEL MICHAEL BARILAN,
MARGHERITA BRUSA

and

AARON CIECHANOVER

OXFORD
UNIVERSITY PRESS

Great Clarendon Street, Oxford, OX2 6DP,
United Kingdom

Oxford University Press is a department of the University of Oxford.
It furthers the University's objective of excellence in research, scholarship,
and education by publishing worldwide. Oxford is a registered trade mark of
Oxford University Press in the UK and in certain other countries

Published in the United States of America by Oxford University Press
198 Madison Avenue, New York, NY 10016, United States of America

British Library Cataloguing in Publication Data

Data available

Library of Congress Control Number: 2021942143

ISBN 978-0-19-886346-5

DOI: 10.1093/oso/9780198863465.001.0001

Printed and bound by
CPI Group (UK) Ltd, Croydon, CR0 4YY

Preface

In 1603, a young Italian naturalist and nobleman, Federico Cesi established the Academy of the Lynxes for Philosophy. It professed mutual support of scholars, especially in relation to freedom of confrontation and debate (Vinti, 2003; Bardiga, 2008). It was probably the first academy to dedicate itself only to research and its dissemination. It was multi-disciplinary too. The Academy's Charter proclaimed that 'The Lynchean Academy desires as its members philosophers who are eager for real knowledge and will give themselves to study of nature, especially mathematics; at the same time it will not neglect the ornaments of elegant literature and philosophy, which, like graceful garments adorn the whole body of science' (Drake, 1966). The Academy's charter, the *Lyncheographum*, required virtuous conduct and thorough self-examination by all members. It was an explicit commitment to seek new discoveries, to document them in the benefit of the public, without causing any harm.

On April 25, 1611, Cesi invited Galileo Galilei to the Academy. In March 1613 the Academy published Galileo's book on the sun spots. Cesi and Galileo chose together additional members and directed research mainly towards astronomy and botany. In 1630, Galileo hoped that Cesi would support the publication of his book on the heliocentric theory, the *Dialogue Concerning Two New Sciences*. Cesi, who was occupied with his attempt to order the whole botanical world in a tabular form, started negotiating with the censor. But he suddenly died in August, before the tasks were accomplished (Freedberg, 2003, pp. 265).

In these very days, the bubonic plague was raging in Lombardy and Veneto. Rome was plague-prone, waiting for the malaria season. Whereas the plague killed half the population of Milano and Parma, Florence was relatively spared. Only twelve percent of its inhabitants died. Immediately after Cesi's death, Galileo decided to flee Rome to his native town, Florence. There, Galileo published his book privately. The chaos resulting from the plague and quarantines prevented him from sending the draft to the Roman censor, satisfying himself with local ecclesiastical authorization instead.

In the spring of 1633, the Roman Inquisition (Congregation of the Index) summoned Galileo to trial. Considering human rationality frail and prone to error, the Church taught that scientific findings contrary to Revelation must not be represented as truths, but only as hypotheticals. Some potentially disruptive ideas, such as Copernicus's heliocentric theory, were excluded from public dissemination. Pope Urban VIII, who had supported Galileo for many years, had to yield to political pressures (Finocchiaro, 2019, ch. 1). Galileo was condemned, forced to retract his theory, and sent to house arrest.

The Lynchaens published a few more works, such as on marine and wood fossils. The books contained magnificent visual representations, but no theoretical explanations. Galileo was not involved anymore, and the Academy dissolved after thirty years of existence.

The Galileo affair left some open questions—what level of certainty may be attached to scientific knowledge? Is it desirable that some social institutions or norms regulate science and the style used by scientists to promote their ideas in public?

Galileo wrote that "philosophy is written in this grand book, the universe . . . but the book cannot be understood unless one first learns to comprehend its language—mathematics" (Drake, 1957, p. 237). With the rise of absolutism, and in the aftermath of the Reformation, knowledge was brought under the administrative competence of the state (Shapin, 1996, p. 129).

The next major scientific revolution, evolutionary theory, inspired some morally problematic ideologies and practices. It became evident that science was not limited to observations and calculations, but it encompassed their representation, reception, and the ethical conclusions scientists and lay people might draw from them. It was not about mere discovery of the truths of nature; but also about critical reflection on the nature of truth (Gingerich, 1982).

In 1801, Abbot and astronomer Feliciano Scarpellini and Francesco Caetani founded the Accademia Caetani, which later assumed the name of the Lynch Academy. In 1847, Pius IX adopted the Lynch Academy as an official institution of the Papal State, the Pontifical Academy of the New Lincei (Cabibbo, 2004, p. 103). In 1870, The Kingdom of Italy took over Rome. In 1874, the statistician, economist and nationalist politician Quintino Sella renewed the Academy as a center for secular science, under the name National Royal Academy of the Lynxes. On March 1, 1874, in his inauguration speech as president of the Academy, Sella proclaimed "There is no higher office to which I can consecrate mayself to other than the development of science in

Rome. I believe it is a duty to nation and humanity" (Guiccioli, 1888, p. 98). Part of the Academy remained in the Vatican.

Like other novel scientific discoveries and theories, Darwin's inspired lively, rich, and creative debates among Catholic theologians; some found it "ennobling", calling people to detect in it new beauties and find sources for optimism (Zahm, 1896, p. 435). But in 1894, the Congregation for the Doctrine of The Faith put an end to the debate by condemning evolutionary theory as erroneous. The Congregation was not worried about erroneous teachings regarding the Heavens, but about errors regarding the human body. The Congregation managed to prevail over Pope Leo XIII, who had entertained more open approaches to science (Brundell, 2001).

Cognizant of the Galileo's affair and the struggle with evolutionary theory, the twentieth-century Church avoided direct confrontation with the racial "sciences". The Church—indeed, all other world religions—did not have a framework for critical reflection on the interplays between science and human values. In 1930, Pope Pius XI did not discuss the validity of eugenics, but protested instead the moral implications of eugenics as a scientific endeavor and social enterprise. He warned against the alliance of state power with the sciences, and their combined efforts to dictate personal values and family life (Lepicard, 1998).

In 1922, Pius XI moved the activity of the Pontifical Academy to Casina Pio IV, a magnificent Renaissance edifice at the heart of the Vatican. Pius XI dedicated efforts to bring modern technologies to the Vatican and to participate in scientific events (Greenwood, 1939).

In the summer of 1936, he re-established a scientific academy that would be independent of a secular [and totalitarian] state. Pius XI transformed the Academy to an international forum. Academicians would be appointed soley by merit of scientific excellence and personal virtues, regardless of religious affiliation or nationality. By answering directly to the Pope, the Academy will be free from worldly patronage (Marini-Bettolo, 1987, p. 4).

These were fateful days. Communism and Racism were promulgated as scientific theories, established on observable truths. Political power dictated scientific research and medical conduct. Totalitarian regimes controlled the territories from which most members of the Academy came from. On October 28, 1936, the Pope changed the Academy's name from *Accademia dei Lincei Filosofi* to the Pontifical Academy of Sciences (Pius XI, 1936).

In March 1937, Pius XI published Encyclicals against Communism (*Divini Redemptoris*) and Nazism (*Mit Brennender Sorge*). He probably realized that ideological abuses of science and technologies call for a different alliance between science, values, and power. Pius XI told the Academy, "The Pontifical Academy represents the Magisterium of Science beside the Magisterium of Faith; the Senate of Science beside the Senate of Church" (Marini-Bettolo, 1987, p. 10).

Pope Pius XII spoke about the Academy as a place of brotherhood in contemplation, study, and engagement in nature (Riccardi, 2003). Religion, philosophy, and science were recognized as independent and interdependet.

In the 1941 annual meeting with the Pontifical Academy of Sciences, Pope Pius XII expressed misgivings about evolutionary theory (O'Leary, 2006, p. 142). Was it possible that science and Enlightenment values were leading humanity into self-destruction? In February 1943, Pius XII addressed the Academy on the subject "The Laws that Govern Nature" (Marini-Battolo 1987, p. 18). In June, the Academy diserpsed. On September 8, Italy surrendered to the Allies. The German army moved in and occupied Rome.

On October 16, 1943, the Nazis raided the Roman Ghetto. Thousands of Jews condemned to deportation found shelter in Castel Gandolfo, the Pope's summer residence, and in more than two hundred other religious houses. It was one of the largest rescue operations during the Holocaust (Loparco, 2012).

After the War, Pius XII laboured hard to preach the independence of clinical judgement from the power of the state and any other external authority. The recognition that religion cannot determine scientific and medical questions, and that state and ideological control of science is even more dangerous, bolstered Pius's conclusion that the personal conscience of clinicians and scientists must be protected and listened to (Pius XII, 1959, pp. 608–18). In a way, he heralded the era of bioethics.

***

Three years ago, Professor Aaron Ciechanover, one of my teachers in medical school and a member of the Pontifical Academy of Sciences, approached me with the idea of a workshop dedicated to the ethical and social implications of 'personalized medicine'. With the academic support of Margherita Brusa, and the generous hospitality of the Academy's Chancellor, Monsignor Marcelo Sánchez Sorondo, its president, Professor Joachim von Braun, and staff, this initiative bore fruit in April, 2019, when distinguished thinkers and scientists deliberated the topic in Casina Pio IV in the Vatican. The event and the related publications are available at the Academy's website.[1]

---

[1] http://www.casinapioiv.va/content/accademia/en/events/2019/medicine.html

This book is an independent creation based on that event, as participants and a few other authors developed their ideas. Much of this work was done during the Covid-19 pandemic. We are grateful to dozens of peers who reviewed the book and its contents. Nicola Wilson, Manhar Ghatora, Sumathy Kumaran, and the team at Oxford University Press have been most accommodating and helpful. We thank the librarians in the Gitter-Smolarz Library of Life Sciences and Medicine, Tel Aviv University. We also thank Dino Maccini and Luca Barberini who granted us permissions to reproduce their art.

We dedicate this book to the caring team in the Covid-19 department MABAM, Tel Aviv Medical Center, and to all Covid-19 caregivers around the world.

Y. M. Barilan

Contents

Author's Biography

Christopher P. Austin earned his MD from Harvard Medical School, and completed clinical training at Massachusetts General Hospital and a research fellowship in genetics at Harvard. Dr Austin directed research and drug development programmes at Merck, with a focus on schizophrenia. From 2016 to 2018, he served as chair of the International Rare Disease Research Consortium (IRDiRC). He later acted as director, National Center for Advancing Translational Sciences, National Institutes of Health, and the National Center for Advancing Translational Sciences (NCATS) at the National Institutes of Health (NIH). He is currently CEO-Partner at Flagship Pioneering.

Yechiel Michael Barilan is a practising physician, expert in internal medicine and Professor of Medical Education at Tel Aviv University's School of Medicine. His research focuses on the interplay between moral philosophy, social history of medicine, and bioethics. He has published *Human Dignity, Human Rights and Responsibility* (MIT Press, 2012) and *Jewish Bioethics* (Cambridge University Press, 2014) and numerous peer-reviewed articles and book chapters.

Giovanni Boniolo is Professor of Philosophy of Science and Medical Humanities, University of Ferrara, Italy; Honorary Ambassador of the Technische Universität München; and President of the Accademia dei Concordi, Rovigo, Italy. He is Co-editor-in-chief of *History and Philosophy of the Life Sciences* and has authored 240 articles, 13 books, and co-edited 12 books. He created and led the PhD programme in philosophy of biology and medicine in the European Institute of Oncology, Milano.

Margherita Brusa earned a PhD in medical humanities from University Complutense, Madrid, and a PhD in paediatric healthcare planning from the University of Padua. In the past ten years, she has conducted research in Tel Aviv University, on the ethics of public health in paediatrics, disaster medicine, and methods of deliberation. She chaired the first ethics committee in the Palestinian authority and participated in ethics committees and ethics education in the USA, Spain, and Italy.

Aaron Ciechanover received his MD from the Hebrew University. During his studies in the Israel Institute of Technology, and under the supervision of Prof. Avram Hershko and in collaboration with Prof. Irwin A. Rose, they discovered the ubiquitin-mediated proteolytic system, for which he was awarded numerous prizes, among them the 2000 Albert Lasker Award for Basic Medical Research and the 2004 Nobel Prize in Chemistry. He is also a member of numerous learned bodies, among them the US National Academies of Sciences and Medicine.

Shlomo Cohen is a Senior Lecturer in the Department of Philosophy at Ben-Gurion University, Israel. He received his PhD in philosophy (*summa cum laude*) and his Medical Doctor degree from Hebrew University, Jerusalem. Shlomo held research fellowships at Columbia University (Fulbright), University of California at Los Angeles, London School of Economics, and New York University. He served as the chairperson of the Institutional Review Board of Ben-Gurion University.

Donna Dickenson is Emeritus Professor of Medical Ethics and Humanities at the University of London. The author of 25 books and more than a hundred academic articles, she has published a formative study of personalized medicine, *Me Medicine vs. We Medicine: Reclaiming Biotechnology for the Common Good* (Columbia University Press, 2013) and a co-edited volume, *Personalised Medicine, Individual Choice and the Common Good* (Cambridge University Press, 2018).

Simon J. Evans, PhD, is a molecular biologist and nutritional scientist. In 2017, he joined the Institute for Systems Biology (ISB) after an academic career at the University of Michigan where his research programme leveraged multi-omics approaches to study dietary patterns that alleviated burden of disease in bipolar disorder. Since at ISB, Dr Evans has focused on the execution of several projects that span Alzheimer disease, diabetes, COVID-19, and others. Dr Evans is currently working with Dr Hood as Chief Science Officer at Phenome Health, a new venture to implement P4 healthcare at the national level.

Joseph J. Fins, MD, MACP, FRCP, is The E. William Davis, Jr MD Professor of Medical Ethics, Chief of the Division of Medical Ethics and Professor of Medicine at Weill Cornell Medical College. He co-directs the Consortium for the Advanced Study of Brain Injury (CASBI) at Weill Cornell Medicine and Rockefeller University and is the Solomon Center Distinguished Scholar in Medicine, Bioethics and the Law and a Visiting Professor of Law at Yale Law School.

Diego Gracia Guillén, MD, PhD, was trained as a psychiatrist. He is Emeritus Professor of History of Medicine and Bioethics. Complutense University of Madrid, Spain. Member of the Spanish Royal Academies of Medicine and Moral and Political Sciences. He is the Director of the Xavier Zubiri Foundation and the Health Sciences Foundation, both located in Madrid.

Leroy Hood got his degrees from Johns Hopkins (MD) and Caltech (PhD). He was a faculty member at Caltech (1970–1992) and University of Washington (1992–2000). He founded the first cross-disciplinary department of biology at the UW and in 2000 co-founded the independent Institute for Systems Biology. He has pioneered important technologies in genomics, proteomics, and single-molecule analyses. He is a member of all three US national academies—science, medicine, and engineering. He has formulated the idea that healthcare should be predictive, preventive, personal, and participatory (P4). He conceptualized deep phenotyping to measure the complexity of individual humans and has pioneered scientific wellness.

Marianne J. Legato, MD, PhD, is an authority in gender-specific medicine. She is now preparing the fourth edition of her textbook, *Principles of Gender-Specific Medicine.* The third edition, as well as her latest book, *The Plasticity of Sex*, have both won Prose Awards from the Association of American Publishers as the best books published in biomedicine in the USA in 2018 and 2021, respectively. She will chair an international symposium in Florence, Italy, in the Spring of 2022, 'Sex, Gender and Epigenetics'.

Farhat Moazam, MD, PhD, a paediatric surgeon, is Professor and Founding Chairperson of the Centre of Biomedical Ethics and Culture (CBEC), SIUT in Pakistan, a WHO-designated Collaborative Center in Bioethics. She has a Doctorate in Religious Studies from the University of Virginia, USA, and is an International Fellow of The Hastings Center, New York. Her work centres on cross-cultural bioethics and its importance in global medical education and healthcare.

Dianne Nicol is a distinguished professor of law and Director of the Centre for Law and Genetics at the University of Tasmania in Australia. She is a fellow of the Australian Academy of Law and the Australian Academy of Health and Medical Sciences and Chair of the Australian National Health and Medical Research Council Embryo Research Licensing Committee.

Roger M. Perlmutter is an MD/PhD graduate of Washington University in St Louis, pursued clinical training at the Massachusetts General Hospital and the University of California, San Francisco, and was a Lecturer in Biology at Caltech before joining the University of Washington, where he was Professor of Medicine and Biochemistry, Chairman of the Department of Immunology, and an Investigator of the Howard Hughes Medical Institute. He previously served as Executive Vice President R&D at Amgen, Inc., and was also Executive Vice President and President, MRL, Merck & Co. He is now CEO of Eikon Therapeutics.

Barbara Prainsack is Professor and Head of Department at the Department of Political Science at the University of Vienna, where she also directs the Centre for the Study of Contemporary Solidarity (CeSCoS), and the interdisciplinary Research Platform 'Governance of Digital Practices'. Her work explores the social, ethical, and regulatory dimensions of genetic and data-driven practices and technologies in biomedicine and forensics.

Nathan D. Price is Chief Scientific Officer of Thorne HealthTech following its acquisition of Onegevity, a AI health intelligence company where he previously served as CEO. In 2019, he was named as one of 10 Emerging Leaders in Health and Medicine by the National Academy of Medicine, and in 2021 he was appointed to the Board on Life Sciences of the National Academies of Sciences, Engineering, and Medicine. He is also (on leave) Professor at the Institute for Systems Biology.

Jenny Reardon is Professor of Sociology and the Founding Director of the Science and Justice Research Center at University of California, Santa Cruz. Her training spans molecular biology, the history of biology, science studies, feminist and critical race studies, and the sociology of science. She is the author of *Race to the Finish: Identity and Governance in an Age of Genomics*

(Princeton University Press, 2005) and *The Postgenomic Condition: Ethics, Justice, Knowledge After the Genome* (Chicago University Press, 2017).

Mehrunisha Suleman is the Director of Medical Ethics and Law Education at the University of Oxford. She holds a medical degree and DPhil in Population Health from the University of Oxford. She also completed a BA in Biomedical Sciences Tripos from the University of Cambridge. Mehrunisha also has an Alimiyyah degree in traditional Islamic studies and was awarded the 2017 National Ibn Sina Muslim News Award for health. Her research interests intersect global health research ethics and clinical ethics.

Henrik Vogt is a medical doctor living in Oslo, Norway, where he has also practised general medicine for several years. He has also studied History and Philosophy. His PhD thesis was titled 'Systems medicine as a theoretical framework for primary care medicine—a critical analysis'. He is currently working as a postdoc at the Centre for Medical Ethics and the Hybrid Technology Hub at the University of Oslo, examining organoids and organ-on-chip technology in conjunction with precision medicine.

Yehoshua Weissinger is a judge in the rabbinic court system of the Chief Rabbinate of Israel. He graduated the MA programne in genetic counselling and is now a PhD candidate in medical ethics at Tel Aviv University School of Medicine. He has published numerous articles in rabbinic journals, mainly in the area of Jewish law and medicine.

James Wilson is Professor of Philosophy at University College London. He is the author of *Philosophy for Public Health and Public Policy: Beyond the Neglectful State* (Oxford University Press, 2021), and has published widely on ethical issues in healthcare and health policy, and on the ownership and governance of ideas and information. His research uses philosophy to address real-world problems, including how these problems reveal gaps and weaknesses in existing philosophical theories. He provides ethics advice to a range of public sector organizations in the UK, and is a member of the National Data Guardian's Panel.

1

Precision and persons in medicine

An introduction

Yechiel Michael Barilan and Margherita Brusa

Personalization and precision

In the past twenty years, 'personalized medicine' and 'precision medicine' have become synonymous. Both terms are *knowledge-oriented*. Medicine is 'personalized' when it is based on information about the person; 'precise' when it is based on information differentiating that person from others. This information is produced by means of informatics ('big data science') of personal biomarkers, especially genetic. Thus, medical information has been taking the lion's share of medical knowledge. Because the data of Artificial Intelligence is "data-sets" and the information elicited comes from highly sophisticated algorithms, some called 'black boxes' (Bleicher, 2017; Rudin and Radin, 2019), the knowledge that ensues is a new kind of knowledge. One technical name given to that knowledge is PAC – Probably Approximately Correct (Haussler, 1990). Is this what policy makers, caregivers, and patients mean when they talk about Precision Medicine?

Information Technology has been pulling medicine in opposing directions. One direction is molecular-level refinement of either existing or imminent diseases, searching for novel cures for at least sub-groups of patients, saving non-responders from side effects and costs of treatment. The other direction is the targeting of whole populations, in search for personalized advice that will improve health and wellness for most.

This book is a journey exploring three dimensions of "personalized/precision medicine" – this new approach to knowledge, the new conceptualizations of health and disease that follows, and their impact on medicine as *praxis*. These epistemological, ontological, and practical developments are dissolving the boundaries separating clinical care, public health, science and technology, bureaucracy, and business. The new kind of knowledge and

Yechiel Michael Barilan and Margherita Brusa, *Precision and persons in medicine* In: *Can precision medicine be personal; Can personalized medicine be precise?*. Edited by: Yechiel Michael Barilan, Margherita Brusa, and Aaron Ciechanover, Oxford University Press. © Oxford University Press 2022. DOI: 10.1093/oso/9780198863465.003.0001

knowledge-making permeate them all, apparently challenging well-established notions of human nature and values.

In Chapter 2, Diego Gracia Guillén surveys the classical systems of classification in biology and medicine, evaluating the metaphysical and epistemological implications of the displacement of medical knowledge by personalized information. In Chapter 4 Giovanni Boniolo brings contemporary philosophy of science to question the tendency to conflate information with knowledge.

The revolution of personalized/precision medicine signals a break from the past, when 'personalization' meant the application of medical knowledge to the care of a specific patient, through special relationships of care, with attention to the psycho-social and even spiritual dimensions. 'Precision' in medicine meant quality of medical instruments, procedures, and drugs relative to standards of engineering.

We may trace the beginning of personalized/precision medicine to the advent of technologies of scientific information about individual humans. Only in the twentieth century did patients' medical records become part of standard care; later, doctors and providers went on to use electronic record keeping, each having their own method of documentation (Reiser, 1978, ch. 10). Personalized medicine evolved from the informatics of clinical trials, whose record keeping has not been designed as an aid to caregivers, but as huge repositories of data ready for statistical analysis. The ethics and law regulating patients' records and databases, created for research and quality control, developed separately from clinical ethics and law. Both have evolved independently from public health records and public registries (Szerter, 2007). Hence, even though 'personalized medicine' is centred on personal information, this information is no longer contained within the enclosed, intimate doctor–patient sphere. Rather, it is stored and processed in diverse forms, spread widely among biomedical institutions and computer networks. The information is "personal," because it is about individual persons; but those persons have less knowledge about its content, spread, and use. Data, especially medical data, have become valuable as such, regardless of any clinical or research context. Data have become a kind of capital. It is collected, stored, and traded for the future benefits it is believed to bring (Sadowski, 2019). Data accumulation and the new conceptualization of 'personalization' peaked in the concept of the "digital twin", a digital treasure trove of all data relevant to a person (e.g. Corral-Acero et al., 2020).

The shift to technologies of personal information in medicine, the expectations invested in this shift, and the social implications of medical dependence on information technology and biotechnology stand at the heart of this book.

Technologies of information and technologies of action

Modernity has given free rein to the scientific search for understanding nature and human nature; the more we know—the better. In contrast to the propitious role of knowledge, action has been the fodder of ethics. If knowledge was feared, it was due to the action it might enable, its manifestation as power. Whereas the regulation of inquiry and knowledge seem medieval, the formative years of bioethics hovered around action. Withholding of life support, allocation of scarce resources, experimentations on humans, and termination of pregnancy are a few notable examples. Personalized medicine is questioning the interplay between the ethics of action and ethics of information, pushing the latter to the fore.

Personalized medicine marks a transition from a kind of scientific insights and clinical ideas to a new vision about the ideals of science and medicine (See Funkenstein, 1986, 19). It abandons abstract diagnostic categories in favour of integrative information on individuals. Medical science has moved from knowledge of human nature to seeking knowledge at the level of individual humans, with special focus on *differences* among individual people. Although genetic knowledge in medicine is the prototype of personal information, the early days of genetics in medicine were not 'personalized'. At that time, a diagnosis of a genetic "defect" revealed something that happened to a person, who was otherwise an ordinary human. The completion of the Human Genome Project, the rise of epigenetics, and the concomitant advent of informatics brought with them the expectation of knowing every person by means of decoding his or her genome while monitoring its 'upper' layers—the transcriptome, proteome, and metabolome. Detection of mutations and variants would not indicate accidents that befall the otherwise 'normal' human genome but would instead be part of the unique genetic identity of the person. This genetic identity has been merged with markers related to metabolism, physiology, nutrition, microbiome, environmental exposure, and lifestyle. In Chapter 10, Joseph Fins shows that it is possible and sometimes critical to refine diagnosis, treatment, and even legal status by means of minute clinical observation, "thick descriptions" (Geertz, 1973, 3–30), and ancillary imaging tests. Fins' discussion of patients with disorders of consciousness creates a vantage point for reflection on the human person.

In Chapter 11, Marianne Legato examines the centrality of sex and awareness of gender in almost every aspect of medical research and care. Gender could be taken as a test case for other macro-factors, which the molecularization of medicine might bypass.

The growing importance of technologies of information alters the meaning of medical interventions. For example, when genetic disease is perceived as an "error" in the natural sequence of nucleotides in the human genome, 'fixing' it fits the paradigm of medicine as a corrective enterprise. When the whole genome is perceived as a unique personal code, the manipulation of sequences is perceived as 'editing' humans. In Chapter 14, Dianne Nicol shows how, despite the universal desire to promote medicine, the laws relevant to 'genome editing' vary markedly among countries. She analyses the regulation of CRISPR and mitochondrial donation, exposing a 'regulatory soup' that mixes diverse laws, and regulations ranging over disparate jurisdictions. The soup's rich content intersects, overlaps and interacts in ways that renders effective governance a key challenge to the legal accommodation of personalized / precision medicine.

Medicine has always aspired to treat patients only by diagnosis (so-called medical need), not allowing external factors, such as poverty and race, to compromise treatment. Complementarily, according to core values of medical ethics, personal information must not leak outside the boundaries of clinical care and must not influence a person's status in society. These ideals have always been subjected to pressures. Today, owing to commercial interests, these pressures seem stronger than ever (ABIM et al., 2002), with the growing dependence of medical research and care on for-profit technological infrastructure, investment markets, and the bureaucratization of healthcare.

The history of medicine is a chain of discoveries and inventions that came to pass in small-scale efforts before they were adopted universally; however, the insights promised by personalized medicine depend on large-scale participation of millions of people to 'share' genetic and other personal data. In any ethical scheme, this participation must be consensual. Hence, lay perceptions of 'personalized medicine' are key to its maturation. These perceptions are especially pressing in relation to predictive biomarkers that lead to non-preventable diseases and complications (Sznajder and Ciechanover, 2012). In Chapter 6, Farhat Moazam argues that people from mid- and low-income countries perceive data collection and processing quite differently from the professionals who work in laboratories, engineers who design the software, regulators, and stockholders. Geographical distances, social backgrounds, gaps in political power, and cultural diversity impact perceptions, the validity of consent and access to benefits. Two chapters in the book introduce religious perspectives. In Chapter 17, Mehrunisha Suleman discusses Sunni Islam, and in Chapter 18, Yehoshua Weissinger and Yechiel Michael Barilan present insights from Jewish law and ethics.

It is the role of the caring physician to mediate the gaps between patients' perceptions and the jargon employed by remote scientists, and to make sense of their situation. In Chapter 12, Shlomo Cohen wonders what role caregivers will have when medicine becomes guided by technologies of information operated, owned, and regulated far beyond the clinical sphere. If we view physician's primary role as a problem solver committed to scientific knowledge rather than communicators of solutions produced by machines, physicians must also trust their own capacities and integrative judgments (see Bolton, 2014). These issues are utmost importance for a medical practice that depends on information flowing to and from individuals in a relationship of care and involving multinational clouds of data. The first pole is committed to agency, relationship, and trust; the second to corporate law, industrial efficiency, and the logic of computation. In Chapter 7, Yechiel Michael Barilan discusses the interplay of these poles, how they might affect people's self-perception, and consequently, their autonomy as patients and as agents in society.

Public health and public policy have been the instruments that arbitrate between social values in relation to health and other sectors of society, especially the market economy. In Chapter 13, James Wilson argues that efficient public health measures need not be precise, and that initiatives of personalized medicine require calibration by the broader contexts of the social determinants of health.

In the years of the COVID pandemic, we are at an increased risk of medical markers determining social status. The classification of people by immune status, 'essential jobs', 'high-risk groups', 'green passes' and even the variants of virus infecting a person, all affect people's capacity to travel, work, and interact socially. The Aristotelian political animal is becoming biologized and molecularized. Human biology is becoming more risk-oriented, political, and commercialized (Braun 2007).

Questions of ownership and distributive justice are central to ethical discourse. Blood donation is commonly considered an act of solidarity, a free gift to the needy. In his research on blood banking, Richard Titmuss shows that a mechanism of voluntary donation is superior to markets in blood (Titmuss, 1997). For additional reasons, it is illegal to sell organs for transplantation. In the era of Information Technology, there is an implicit expectation to freely share personal medical data, for the benefit of all, with the promise of "anonymization".

Personal data is personal because it stores information about the person from whom the data originates. This unique situation renders information 'sharing' inextricable from social relationships and their ongoing ethical obligations (Soo-Jin Lee, 2021). Some authors advocate continuous personal control over one's own information (Lanier, 2013; Topol, 2019). Inspired by Annette Weiner's theory about material gifts, one may suggest that personal information is a kind of 'inalienable possession', a good that—even when given away—is given within a broader framework of human relationship. Ultimately the gift returns somehow to the owner, closing a circle of benefit and recognition (Weiner, 1992). This entails something beyond the mere power to retract shared information or to monetize it, as Topol and Lanier advocate. It includes a right to benefit from its products as well as a capacity to participate in regulating the process.

This is borne out in Chapter 9, where Christopher Austin narrates the story of 'rare disease medicine', which relies on the mutual support and grassroots activity of people worldwide. Even though they are scattered all over the globe, carriers of rare genetic mutations depend on each other for genetic data, participation in clinical trials, and other forms of collaborations. Personalization depends on solidarity.

In the beginning of the twentieth century, especially with the infant welfare movement, universal surveys of babies became the standard of good public health and childcare. In the 1960s, the introduction of newborn screening initiated mass scrutiny of babies at the molecular level in search of one rare inborn error of metabolism. In Chapter 8, Margherita Brusa and Donna Dickenson discuss the proposed transition of newborn screening from a public health enterprise, aimed at the prevention of a few catastrophic diseases, to universal genetic screening of newborns, questioning whether such an operation is in the best interest of the child.

Historically, the term personalized / precision medicine appeared in relation to diagnosis refinement and variability of patients' responses to drugs (pharmacogenetics). For example, patients with lung cancer are becoming ever more subdivided by a growing number of molecular markers (Ciriello Pothier, 2017, p. 17). In his 2015 State of the Nation Address, President Obama announced the 'Personalized Medicine Initiative', bringing the concept into the forefront of research and funding.

Rare disease medicine and newborn screening embody concentrated efforts directed at a few but seriously afflicted people, trying to capture the source of plight at the molecular level. Whereas rare disease medicine and newborn screening search a few biologically unique persons among the many; personalized medicine has been moving to search the biologically unique in every person.

In Chapter 3, Leroy Hood, Nathan Price and Simon Evans lay out a far-reaching conceptualization of personalized medicine, 'P4' (Personal, Predictive, Preventive, and Participatory). Its scientific approach is inspired by system biology. The operation is reminiscent of newborn screening by its universal appeal, dependence on mass participation, and prevention that is based on prediction. P4's meticulous analysis of every participant patient renders the operation like rare disease medicine too. Whereas rare disease medicine and newborn screening seek to significantly improve the lives of the few, P4 aspires to accrue numerous health benefits to almost everybody. In Hood, Price and Evans' scheme, implementation of P4 requires personal counselling and coaching by dedicated professionals. In Chapter 15, Barbara Prainsack asks whether minorities and underserved groups might be side-lined by bias in the datasets and algorithms. These minorities, especially the poor, might not have the resources to afford counselling about their ongoing health-information.

In line with the biopsychosocial model, P4 personalized medicine works on the assumption that the more information is better, that "truth is achieved by adding more and more perspectives, getting closer and closer to a highly complex reality. This is common sense, perhaps, but not scientific sense" (Ghaemi 2009). These words of criticism were written in relation to the biopsychosocial model. It remains to be seen whether P4 and similar operations manage to bring forth valid science on the commonsense assumption that more data is more information, and more information is better healthcare.

Personalized medicine and the allure of magic

Whereas people may be divided into those who believe in 'science' and those who believe in 'magic', in his book *Witchcraft Among the Azande*, anthropologist Edward Evans-Pritchard theorizes that both modes of thinking provide complementary ways of understanding reality (Evans-Pritchard, 1937). Observable phenomena allow us to infer patterns. People can tell that storms generate lightning, and that lightning ignite dangerous fires. The study of lightning and wildfires may become deeper and more comprehensive, but it will not answer the personal question 'why did the lightning hit my house and not my neighbour's?' According to Evans-Pritchard, the personal question invites magical explanations. A malicious spell cast on a person might explain why, of all huts in the village, only his went ablaze. Science will explain the arrival of the storm and the incendiary power of lightning; magic will account for their apparently selective effects at the personal level. Science is about generalizations and principles. Magic adds this kind of precision. The precision is always personal. Only witchcraft can pinpoint the person and the timing.

The moto "right drug, right dose, right time, right patient, right route" originated in the literature on patients' safety (Grissinger 2010). Personalized medicine aims to know the right dosage of the right intervention, at the right time for every patient. Can science break its own epistemological and metaphysical boundaries and perform tasks nobody believed possible? Will personalized medicine cure all disease? Does personalized medicine bode a new medical utopia? Are the promises of personalization and precision fully consistent with each other?

The interchangeable terms 'personalized medicine'/'precision medicine' invoke the archaic desire to know 'why me?', and to connect *knowledge of the world* with *self-knowledge*. Even though 'knowledge of the world' is a noble title for science, in the era of scientific medicine, the production of clinical truths has passed a few phases, among them expert consensus, the 'bio-psycho-social' model, and 'evidence-based medicine'. In Chapter 5, Henrik Vogt shows conceptual incompatibilities between the art of personalization with the generalizability of science. Today, clinicians do not have to choose between evidence-based practice and holistic care; nor is expert consensus outmoded. Competent medicine is aware of the powers and limitations of each approach. Enthusiastic embrace of 'personalized medicine', as if it is a comprehensive alternative to other forms of medicine, runs the risk of forgoing both its promises and limitations, as well as the risk of catering to human narcissism and its fascination with magic.

The conflation of information with knowledge and the allure of holism might boost people's tendency to approach technology as a kind of magic and to go for solutions that imitate technology, even solutions to technology's harms (Ellul, 1981, p. 48). Following Hannah Arendt, in Chapter 16, Jenny Reardon argues that displacement of reflective thinking by engineering mindsets accelerates such processes. Simplistic reception of personalized medicine might generate self-knowledge and self-esteem that are predicated by biotechnology and information technology.

There is a deep-seated craving to know the answer 'why me?', 'why now?'. In opposition to cultures centred on magic, the art of medicine has focused on tailoring scientific knowledge to patients' needs. The artful doctor would not necessarily tell the patient why, unlike others with identical risk profile, she had a 'heart attack'. The doctor's art will allow her to emphatically grasp broader aspects of the patient's condition, to construct meaning, to communicate better, and thus inspire compliance and healing (Solomon, 2015, ch. 8). The personal aspect of medicine has been about the qualitative bridge that connects the impersonal science with the individual patient as a subject. Personalized medicine might add layers to traditional practice and may even transform it; but it might also alter humanity's notion of what "personal", "person" and "personalization" mean. It does not yet offer an approach to diverse kinds of suffering and debility, which still eludes scientific labels and biomarkers. Titles such as chronic pain syndromes, fibromyalgia, chronic fatigue syndrome, as well as many psychiatric disorders may call for different diagnostic and therapeutic approaches altogether.

The book concludes with Roger Perlmutter's review of personalized medicine within the history of medicine as a process that reduces morbidity and mortality. In Chapter 19, Perlmutter argues that proper evaluation of personalized medicine will be better achieved by focusing on its historical relationship with other social, scientific, and technological developments.

When 'personalized medicine' talks about the 'right patient', it refers to the patient who is likely to respond to treatment. We should remember that many patients are 'right' but do not receive care solely because of financial, cultural, and political barriers. Many would not have been sick in the first place, had they not been exposed to environmental, occupational, and other human-made hazards, let alone violence and deprivation. Even if personalized medicine delivers all of its promises, medicine's global mission is to reach every sick person, anywhere in the world, anytime he or she needs care.

2

The historical background of personalized medicine

Diego Gracia Guillén

Specification, individuation, personalization

Something new is happening. We are abandoning the classical way of understanding disease as a discrete process. The ideal of medicine as a science has been knowledge of the causes, symptoms, and treatments of diseases. Scientific medicine has been predicated by the idea of specificity. As a consequence, individual patients were only 'cases', with many variable and confounding circumstances added to the pathognomonic symptoms of diseases. Disease is a universal category, whereas particular manifestations differ from one individual patient to another.

This was the traditional way of medical thinking. Now, we are shifting to a new logic, in which disease is conceptualized as an individualized phenomenon rather than as an abstract entity that manifests itself in diverse individuals. This is the novelty that has been called 'personalized medicine'. It is a new way of beholding disease. But at least in philosophy, 'the person' is not the same as the individual. The goal should be referred to as 'individualized medicine', 'genomic medicine', or 'precision medicine'. Personalizing medicine is no less complex an objective, and remains a completely open question.

Scientia 'de universalibus'

Throughout history, medicine has passed two major phases. The critical point between them appeared when Greek philosophers defined 'science' (*episteme*) as an apodictic knowledge of universal truths. *De particularibus non est escientia*. There is no science of particular cases, the old saying went, because 'that which is individual and has the character of a unit is

Diego Gracia Guillén, *The historical background of personalized medicine* In: *Can precision medicine be personal; Can personalized medicine be precise?*. Edited by: Yechiel Michael Barilan, Margherita Brusa, and Aaron Ciechanover, Oxford University Press. © Oxford University Press 2022. DOI: 10.1093/oso/9780198863465.003.0002

never predicable of a subject' (Aristotle, *Categories* 2: 1 b 5). The medieval Aristotelian philosophers coined the expression: *individuum est ineffabile*, the individual is ineffable and therefore there is no scientific knowledge of it. This was the historical moment in which the archaic 'empiricist medicine' began to be substituted by 'scientific' medicine. The Aristotelian logic stated that essence is neither individual nor personal but 'specific', and that a species is the result of combining a *genus*, 'genre' ('mineral', 'vegetal', 'animal') with a *difference* or 'specific difference' ('with feet', 'two-footed', 'winged', 'aquatic', as differences inside the genre 'animal'). The most typical example is the case of the human being. We can give him or her external and descriptive definitions, for instance, a two-footed animal. But the only possible substantive definition of the human species is 'rational animal'. There are few animals with two legs; but only humans are rational. The other differences that mark one human being from another are only 'accidental', not essential.

What can be said about an object? The classical praedicabilia

Praedicabilia are the things that can be said about an object; the things or words that predicate it. Aristotle (*Topics* I 5: 102 a–b) identified four *praedicabilia*: definition ('*Horos* is a phrase signifying a thing's essence'); genus ('*Genos* is what is predicated in the category of essence of a number of things exhibiting differences in kind'); property ('*Idios*, a property, is a predicate which does not indicate the essence of a thing, but yet belongs to that thing alone, and is predicated convertibly of it'); and accident ('*Symbebekos* is something which, though it is none of the foregoing, i.e. neither a definition nor a property nor a genus yet belongs to the thing'). Porphyry added a fifth, *diaphora*, *differentia*, or difference (Porphyry, 1853, p. 618). The scholastic classification, obtained from Boethius' version of the *Isagoge*, modified Aristotle's by substituting definition (*horos*) by species (*eidos*), because an essential definition is necessarily specific (Boethius, 1847, p. 37). Species, as a consequence, is the result of adding to the genus one specific difference.

The theoretical knowledge of species is the proper goal of science. The practical consequences are managed by artisans through the analysis of their properties (the signs that appear necessarily when the species is present), and their accidents (signs that may or may not be present when the species is present).

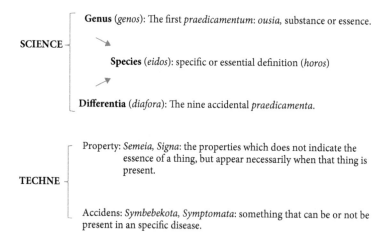

SCIENCE

Genus (*genos*): The first *praedicamentum*: *ousia,* substance or essence.

Species (*eidos*): specific or essential definition (*horos*)

Differentia (*diafora*): The nine accidental *praedicamenta.*

TECHNE

Property: *Semeia, Signa*: the properties which does not indicate the essence of a thing, but appear necessarily when that thing is present.

Accidens: *Symbebekota, Symptomata*: something that can be or not be present in an specific disease.

The Porphyrian tree

Following the Aristotelian description of *categoremata* or *praedicabilia,* Porphyry, a Neoplatonic philosopher (third century CE), designed his *scala praedicamentalis* (Porphyry, 1853). Medieval logicians represented these ideas by means of tree-like diagrams. The diagrams were comprised of dichotomous divisions, which indicate that a species is defined by a *genus* and a *differentia.* Biforcation continues until the lowest definable species is reached. At this stage, the metaphysical analysis has reached the ineffability of the individual (Figure 2.1 from Boethius, 1847, pp. 41–2; copyright in public domain).

Boethius's sixth century Latin translation of Porphyry's text became the standard approach to logic throughout the Middle Ages, remaining in textbooks in schools and universities until the late nineteenth century.

This *scala praedicamentalis of* Porphyry was the model of structuring and organizing the *scientia naturalis* throughout the Middle Ages. The knowledge of nature was understood as the study of the 'scale of beings' or *scala naturae.* Linnaeus's system of nature was also conceived in the same way, as declared in the title of his book: *A general system of nature through the three grand kingdoms of animals, vegetables, and minerals; systematically divided into their several classes, orders, genera, species, and varieties.* The steps of the new scale were then: classes, orders, genera, species, and varieties. Linné pointed: 'In this arrangement, the *classes* and *orders* are arbitrary; the *genera* and *species* are natural. All true knowledge refers to the species, all solid knowledge to the genus' (Linné, 1802, I, p. 3).

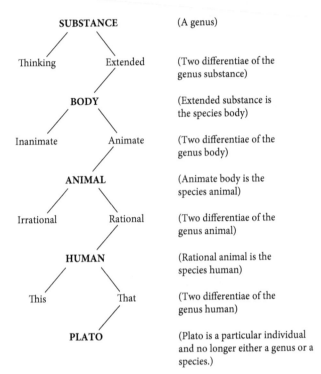

Figure 2.1 From the scala praedicamentalis to the scala naturae.

The case of human beings is particularly instructive. They are members of the class Mammalia, the order Primates, the genera *Homo*, and the species *sapiens*. This last is the main concept, because 'all true knowledge refers to the species'. The name of a species is binary. Hence, humans who belong to the genus *Homo*, are called *Homo sapiens* because rationality (*rationalitas*) differentiates them from all other species.[1] Following this logic, Linné thought that there was another species inside the genus *Homo*, which he called *monstrosus*. There was a variety of monsters: Mountainer, Patagonian, Hottentot, American, Chinese, and Canadian. Inside the *Homo sapiens* species, Linné described different varieties: Wild Man, Rubescens American, Albus European, Brown Asiatic, and Niger African (Barilan, 2012; Livingstone, 2008).

[1] The expresion "Homo sapiens" was used by Linné in *Systema naturae*, tomus I, ed. 1758, p. 20.

Medicine and specificity

Greek physicians, especially Galen of Pergamum, applied this method of classification to diseases, arranging them in different 'genera' (*gene*) and identifying the 'differences' (*diaphorai*) necessary to distinguish all the 'species' (*eide*) of disease inside each genus (*gene*) (Galen, 1964, I, p. 259; VII, pp. 42–3). Using this method, Galen defined disease (*nosos*) as 'a state contrary to nature [*genus*] by which an action is damaged primarily [*differentia*]' (Galen, 1964, VII, p. 43). Nosology (*nosologia*) is the science focused on the study of the essence of diseases, called *noson eidos*, and in Latin *species morbosa*. On the other hand, the goal of pathology (*pathologia*) is the study of individual patients in which the disease is present, analysing their permanent signs and its inconstant symptoms. Pathology, therefore, cannot be a scientific knowledge (*episteme iatrike*), but only an art (*techne iatrike*), centred in the 'descriptive' knowledge of individual patients through durable signs and accidental symptoms.

NOSOLOGY
- **Genus**: Liver (*sykon* in Greek, *ficus* in Latin, is the origin of *ficatum,* liver in the common language. In the medical language, liver was called *hepar* both in Greek and Latin)
- **Differentia**: Damaged in *poion*, or quality: hardened (*skirros,* hard, induration, hardened swelling or tumor).
- **Species**: Cirrhosis (liver hardened)

PATHOLOGY
- **Signum**: *Clinical signs*: Liver hardened to palpation
- **Accidens**: *Clinical symptoms*: hemorrhage

The medical school of 'nosographers', with François B. de Sauvages (1763) and Philippe Pinel (1798) among many others, applied to human diseases the same method used by Linné in his work on plants and animals. The model was based on the Porphyrian tree.

The chain of specificity

All the efforts of western medicine, from the ancient Greece until half a century ago, were oriented to identify the 'chain of specificity'. Every disease has a 'specific cause', called aetiology, which produces specific disease (*species morbosa*), which medicine should revert through specific treatment.

SPECIFIC AETIOLOGY ➡ SPECIFIC NOSOLOGY ➡ SPECIFIC THERAPEUTICS

From specificity to individuality

Following the example of Plato, science was classically understood as an apodictic knowledge about the *quidditas* or essence of things. Philosophers identified 'essence' with 'quidditas' or 'species' (Zubiri, 1980, pp. 218–19). Consequently, the 'empirical' or 'experimental' knowledge cannot be conceived as scientific, because it is always particular.

The medieval 'Nominalist' school criticized the Neoplatonic paradigm that divides reality into distinct essences, arguing that the world is an assortment of unique events.

In the seventeenth century, nominalism facilitated the emergence of experimental science, which is the knowledge of nature as direct observation and measurements.

This revolution culminated in the twentieth century when philosophers identified essence with the physical reality of things, and not with logical entities. 'We find ourselves referred back from the quiddity, as predicate, to physical reality. It is not a question, as some scholastics already thought, of amplifying the perimeter of quidditative predication with individualizing predicates, but of abandoning all predication in order to go to the thing in its physical reality' (Zubiri, 1980, p. 221). Reality is comprised of individual items and events. The essence is what 'confers on it its minimal and ultimate character' (Zubiri, 1980, p. 221). Therefore, essence, or 'constitutive essence', is the minimum quantity of elements or properties a thing needs to be what it is (Zubiri, 1980, pp. 224–225, 228–229).

The old idea that specificity is the only way of knowing the essence or true nature of a reality has been strongly criticized from the philosophical point of view, but it is also incompatible with direct observations. Aristotle, the author of the most successful theory of specificity in history, described cases in which his theory didn't work, referring to such exceptions as 'monsters' (*terata*) (Foresman, 2013, pp. 16–25). In the case of human species, monstrosities 'do not at any rate resemble a human being' (Aristotle, Generation of Animals, 767b 4–5). Today we may explain Aristotelian 'monstrosities' as mutations, hybridizations, epigenetic changes, and other processes that alter the genome and its expression. Aristotle may be seen as an ancestral influence toward the new scientific concept of hybridization (Generation of Animals, 746 a 29-b). In evolutionary biology, hybridization is one of the most important arguments against the idea of 'specificity'.

Three years before the publication of *The Origin of Species*, Charles Darwin wrote that: to speak of 'species' is 'trying to define the indefinable' (Sowa, 2016, pp. 5–13). The discovery of Neanderthal, Denisovan, and even viral genome in present-day human genes calls into question the essence theory of the species, at least its genetic terms (Reich, 2018).

Something similar has happened in medicine, namely a shift from the constitution of diseases to their concrete manifestations. *Species morbosae* are mere abstractions, heuristics. The only real disease is that which is suffered by each individual. Genes, *prima facie* taken as specific elements, react differently depending on environmental conditions, epigenetics, mutations, and many other factors. The consequence is that medicine is aware of individual susceptibilities to drugs that hitherto were taken as 'specific' treatments. Medicine is changing gradually from the old ideal of 'specificity' to the new one of 'individuality'. This has been the conceptual origin of what has been called 'personalized medicine', but it deserves the name 'individualized medicine'.

From 'disease' to 'disorder'

One consequence of these developments in thinking about essences and diseases has been the growing use of the term 'disorder' at the expense of 'disease' (McNally, 2011). This transition began in psychiatry, when it was recognized that many symptoms of distress or discomfort that alter the normal functioning of human beings in their environment, but without the characteristics of mental diseases, can be medicalized and treated. But this process is becoming ever more evident in other areas of medicine—for example, categories such as 'medicalization', 'wish-fulfilling medicine' and 'pharmaceuticalization' are becoming common (Blasco-Fontecilla, 2014; Meneu, 2018). In fact, the *Diagnostic and Statistical Manual* used in psychiatry no longer uses the expression 'mental disease', which has been substituted by 'mental disorder', adding that the definition of these disorders depends on 'cultural, social, and familial norms and values' (American Psychiatric Association, 2013, p. 14). Whereas the 'bio-psycho-social' model of disease called for cultural awareness, the new approach to disorders incorporates cultural factors in the very conceptualization of diagnostic categories (Lewis-Fernández and Aggarwal, 2013; Warren, 2013; Bredström, 2019). According to the American Psychiatric Association,

A mental disorder is a syndrome characterized by clinically significant disturbance in an individual's cognition, emotion, regulation or behavior that reflects a dysfunction in the psychological, biological, or developmental processes underlying mental functioning. Mental disorders are usually associated with significant distress or disability in social, occupational, or other important activities. (American Psychiatric Association, 2013, p. 20).

The consequence is that human health—not just mental disease—can no longer be defined only by biological facts (for instance, by genetic traits), but needs also to include communal and personal values. 'The boundaries between normality and pathology vary across cultures for specific types of behaviors. Thresholds of tolerance for specific symptoms or behaviors differ across cultures, social settings, and families. The judgment that a given behavior is abnormal and requires clinical attention depends on cultural norms that are internalized by the individual and applied by others around them, including family members and clinicians' (American Psychiatric Association, 2013, p. 14).

Whereas the classical idea of disease was strictly biological, this new approach based on the idea of disorder gives room to the new category of values. This novelty is also present in the manual of the *International Classification of Diseases* (ICD-11) of the World Health Organization. The word 'disease' is used with the classical meaning it has in medicine, whereas the word 'disorder' seems to be used when personal, social, and cultural values play a more pathogenic role. The consequence is that the shift from the old specificity to the new individuation is not only related to the capacity of modern biology to identify genetic differences and susceptibilities, but also to the personal, social, and cultural varieties in values.

The consequences of this new approach are unpredictable. One may suggest that the incorporation of values in medical categories may entail the selection of traits that are culturally considered 'positive'. Is personalized medicine destined to become a new kind of eugenics? Will it transform medicine from a biologically limited focus on a species-typical functionality (Boorse, 1997; Giroux, 2015) to a pluralism of individual cases and preferences? Modern sociology has always claimed that even biological medical judgements are not 'value free'. The explicit incorporation of value in the conceptualization of diseases (or: lack of wellness) might threaten to medicalize all dimensions of cultural and political life.

Looking for a better name: precision medicine

Perhaps, we should distinguish between 'individualized medicine' and 'personalized medicine', understanding by the first the use of genomics to improve our capability of diagnosing and curing diseases, while avoiding 'enhancement' of human nature. Philosophy has stressed that personality cannot be identified with individuality. The use of the word 'person' or 'personality' indicates an integrative conceptualization of the human being, placing consciousness and its cultural, moral, and religious manifestation at the centre. Genuine 'personalization' of medicine should incorporate personality, a goal that seems quite remote.

Because 'personalized medicine' is not an appropriate name, new candidates have appeared, like 'genomic medicine', and 'precision medicine'. This last has been introduced after the publication by the National Academy of Sciences of the consensus study report of the National Research Council Committee on A Framework for Developing a New Taxonomy of Disease, entitled *Toward Precision Medicine: Building a Knowledge Network for Biomedical Research and a New Taxonomy of Disease* (2011). The report defines precision medicine as 'the tailoring of medical treatment to the individual characteristics of each patient'. This is now possible because the new molecular techniques are capable of identifying the particular predispositions to certain diseases, and the individual susceptibilities to specific drugs and products, making possible to target medicine to susceptibilities and statistical profiles rather than embracing the 'person' directly.

The report by the National Academy of Sciences justifies the use of the expression 'precision medicine' instead of 'personalized medicine'. It says: 'Although the term "personalized medicine" is also used to convey this meaning, that term is sometimes misinterpreted as implying that unique treatments can be designed for each individual. For this reason, the Committee thinks that the term 'precision medicine' is preferable to 'personalized medicine' to convey the meaning intended in this report.'

The Mission Statement of the Precision Medicine Initiative reads: 'To enable a new era of medicine through research, technology, and policies that empower patients, researchers, and providers to work together toward development of individualized care' (Obama, 2015). One of the first consequences of this movement has been the '*All of Us* Research Program' initiative, presented as the effort to gather data from at least one million people to accelerate research and improve health. By taking into account individual differences in lifestyle, environment, and biology, researchers will uncover paths toward delivering precision medicine.

3

What 21st century medicine should be—history, vision, implementation, and opportunities

Leroy Hood, Nathan D. Price, and Simon J. Evans

Introduction

Healthcare today faces four major challenges:

1. Quality: an effective healthcare system must address the complexity and uniqueness of each individual to achieve quality outcomes. The USA ranks at or near the bottom of the top twenty developed countries in this regard.

2. Cost: healthcare costs globally continue to rise in comparison with other economic sectors (Organisation for Economic Co-operation and Development, 2020). In the USA, healthcare spending climbed from 5% gross domestic product in 1960 to nearly 18% in 2019 (Hartman et al., 2020) without commensurate improvements in health outcomes. The USA spends close to three times as much on healthcare as the next most expensive developed country and achieves far less (Emanuel, 2020).

3. Ageing: the percentage of the global population aged >60 years increased from 16% in 1980 to 24% in 2015 with projections to reach 34% by 2050. Clearly our ageing population suffers greatly from the typical decline in both body and brain health as we age.

4. Chronic disease: over the past two decades, non-communicable chronic diseases have risen from 43% to 62% of the total global disease burden, and the growth of the most common chronic diseases—diabetes, obesity, and neurodegenerative diseases such as Alzheimer disease—continues to grow explosively.

Leroy Hood, Nathan D. Price, and Simon J. Evans, *What 21st century medicine should be—history, vision, implementation, and opportunities* In: *Can precision medicine be personal; Can personalized medicine be precise?*. Edited by: Yechiel Michael Barilan, Margherita Brusa, and Aaron Ciechanover, Oxford University Press. © Oxford University Press 2022. DOI: 10.1093/oso/9780198863465.003.0003

Each of these major challenges can be addressed by a systems (global or holistic) approach to a healthcare that is predictive, preventive, personalized, and participatory (P4). We believe that 21st century medicine, fuelled by a P4 approach, will provide powerful strategies to deal with each of these four challenges. These strategies will motivate, and be enabled by, a shift in healthcare from a fee-for-service to a value-based payment paradigm for medicine. This chapter provides a detailed discussion of the history, vision, and future implementation of what 21st century medicine should be.

Seven paradigm changes—a historical perspective

Leroy Hood has been part to seven paradigm changes facilitated the transition from modern medicine to p4 medicine (Hewick et al., 1981; Hood, 2008, 2019). These were driven by a conviction that humans are terribly complex creatures and that, in order to assess this complexity with regard to wellness and disease, one must generate significant multidimensional data on each individual. This conviction catalysed a series of seven paradigm changes starting in 1970 when Hood began his career at Caltech, focused on pioneering technologies to decipher human complexity. These seven paradigm changes dealt with biological complexity and framed our vision of what 21st century medicine should be.

Bringing engineering to biology

Beginning in the early 1970s and for the next thirty-five years the Hood lab worked on developing instruments that enabled analysis of DNA and proteins in new ways (Hunkapiller et al., 1984; Kent et al., 1984; Smith et al., 1986; Horvath et al., 1987; Blanchard et al., 1996; Geiss et al., 2008). These instruments allowed us to develop a powerful new integrative strategy for cloning genes. In 1981, Hood co-founded Applied Biosystems to bring the automated technologies (DNA sequencer, protein sequencer, DNA synthesizer, and peptide synthesizer) to the scientific community. Such instruments created the foundational technologies for biotechnology, allowing us to automate, integrate, increase the throughput, and in some cases even miniaturize critical chemistries.

The Human Genome Project

The second paradigm change was the realization of the Human Genome Project. As the developer of the automated DNA sequencer, Hood was invited to the first meeting discussing the genome project at Santa Cruz in the spring of 1985, and was asked, along with eleven others, to pass judgement on whether the Human Genome Project was a good idea. The decision was that it was feasible, but the invited experts were split six to six on whether it was an idea worth pursuing. In the mid-1980s, perhaps 80% of biologists in the USA were opposed to the Human Genome Project, as were the National Institutes of Health. A major counter-argument was that it represented a large, coordinated 'big science' effort that could conceivably take resources away from single-investigator-centric small science—which was the essence of biology at that time. By the time this project became a reality in 1990, attitudes had started to shift due to the demonstrations of the power of automated DNA sequencing, and the project secured new funding in addition to the already-existing governmental allocations, meaning that it did not compete with the small science efforts. Quite to the contrary, the Human Genome Project and the other genome projects to follow would prove a boon that have since supported an immense number of investigator-led projects. The human genome sequence was the second paradigm change, for it gave us the ability to correlate human genetic variants with both wellness and disease phenotypes— a central pillar of what we will argue later should be a part of 21st century medicine.

Cross-disciplinary biology

The third paradigm change was the formalization of cross-disciplinary biology. The automated DNA sequencer required the integration of biology, chemistry, engineering, computer science, math, and molecular biology. Hood proposed at Caltech to create a second biology department that would be an applied, cross-disciplinary department, but failed to persuade his biology colleagues that this was a good idea. In 1992, Bill Gates made it possible for Hood to realize this vision at the University of Washington. There, gathered under one roof all the flavours of scientists required for diverse biological technology projects—biologists, chemists, computer scientists, engineers, mathematicians, physicians, and physicists. This approach drove discovery in science, and it drove innovation. This cross-disciplinary approach led to the development of a series of technologies and strategies that were critical for the next paradigm change. In 2000, Hood resigned from the University of Washington to start the non-profit Institute for Systems Biology (ISB).

Systems biology—how to deal with biological complexity

The fourth paradigm change was the formalization of the new discipline of systems biology, building it on a cross-disciplinary platform. The ISB was the first of its kind (Hood et al., 2008). Systems biology is a holistic, global, integrative, dynamic, and cross-disciplinary approach to biology that differed substantially from the classical reductionistic biology of that time, which studied biological systems one gene or protein at a time (Hood et al., 2004). We applied systems approaches to major problems in biology and disease (Ideker et al., 2001b; Price et al., 2007; Hwang et al., 2009; Roach et al., 2010). We also triggered the fifth paradigm change by defining the fundamental concepts leading to 21st century healthcare. There are now hundreds of centres, institutes, etc., of systems biology across the world. Systems biology is a major approach to dealing with biological complexity widely accepted across the biological sciences.

Conceptualization of 21st century medicine

The fifth paradigm change was the conceptualization of 21st century medicine. We called the systems approach to study disease 'systems medicine', which includes three things: the use of longitudinal, deep phenotyping to characterize the complexities of disease; the use of network biology to understand the mechanistic underpinnings of various types of diseases, as well as promoting the ability to discover biomarkers and candidate drug targets; and the creation and integration of new technologies, computational platforms and diverse data types to deal with the complexities of biology and disease (Weston and Hood, 2004). The experience with systems biology led to a shift in thinking about the nature of healthcare itself—namely, the idea that healthcare needed to be predictive, preventive, personalized, and participatory—forming the P4 medicine or healthcare framework (Hood and Friend, 2011; Hood and Flores, 2012). The essence of P4 healthcare is the understanding that there are two domains in healthcare: wellness and disease (Hood and Price, 2014). At that time (early 2000s), wellness was virtually ignored in most healthcare systems, with the vast majority of resources dedicated to reacting to disease onset. The conceptualization of systems medicine and P4 healthcare lies at the very heart of what 21st century medicine should be.

Scientific (or quantitative) wellness

The sixth paradigm change was the pioneering of scientific wellness. In 2014, we put together a pilot project to explore the potential of quantifying wellness. The research was based on genome analysis and longitudinal deep phenotyping (measurement of thousands of blood analytes, gut microbiome composition, and lifestyle behaviours, e.g. sleep, physical activity, and dietary intake, etc.) of 108 individuals for a period of nine months (clinical chemistries, proteins, and metabolites from the blood; microbiomes from the gut; and digital quantified health measurements). Blood and stool samples were taken every three months and data-driven personalized feedback was given to individuals through health coaches trained in nutrition, exercise, and psychology. We had two spectacular results (Price et al., 2017). The first was that we could strikingly change the wellness of individuals with this approach as measured by classical health metrics; the second was that the data clouds we generated could be analysed to identify new approaches and insights into both biology and medicine (Earls et al., 2019; Wilmanski et al., 2019; Zubair et al., 2019; Wainberg et al., 2020).

Bringing 21st century medicine to the US healthcare system

The final paradigm change began in 2016, when Hood was recruited by the CEO of Providence St Joseph Health (PSJH) as Chief Science Officer. PSJH is the third largest non-profit healthcare system in the USA with fifty-one hospitals along the west coast, seeing 10 million patients per year. His mission was to bring systems thinking to PSJH clinical research and scientific wellness to their clinics. The decision was also made to affiliate ISB with PSJH to further cement this systems approach to medicine. There ISB initiated a new project that began in 2020 to characterize individuals by genome and longitudinal phenome analyses at a scale of 1 million patients over a five-year period—the PSJH million person genome/phenome project.

These seven novel approaches reshuffled the disciplinary toolkit of both basic science and clinical medicine, bringing engineering, computer science, and molecular biology to the 'bedside', which is actually to the ordinary lives of people whose personal interaction with these technologies is expected to provide unprecedented monitoring of their health and disease processes.

Contrasting contemporary or 20th century medicine and our vision of 21st century medicine

Before the seven paradigm changes, the healthcare system was focused on curative medicine—to find and fix disease by scientific methods. The 1910 Flexner report of the Carnegie Institute reviewed the then-existing 155 trade medical schools in the USA and Canada and argued that science should drive patient treatment and physician training (Flexner, 1910). Johns Hopkins and a few other bold medical schools jumped on this concept, which greatly improved the quality of physicians licensed to practice medicine across the country. Independently of the Flexner report, the twentieth century saw major improvement in public health—water sanitation, immunization, enrichment of salt with iodine, environmental clean-up, health education regarding the risks of smoking, alcohol and obesity and lack of exercise.

However, 20th century medicine has been much less successful treating chronic diseases arising from multifactorial causes, many of which were intensified by contemporary lifestyle changes and environmental exposures of the industrial age.

We argue that a second major paradigm shift in medicine is happening now, building upon the successes of 20th century medicine, and is framed by the seven paradigm shifts discussed earlier by transforming how we think about the complexities of biology and disease. The health needs of the 21st century are vastly different from those at the start of the 20th century, with 86% of total healthcare dollars in the USA now spent on chronic disease (Buttorff et al., 2017; UCL Institute for Global Prosperity, 2017) and comparatively little on most infectious diseases. There remain of course very significant infectious global health challenges, which we will not address herein as well as infrequent periodic events such as the global COVID-19 pandemic. New approaches are desperately needed, and they are arising from taking systems approaches to healthcare.

Systems approaches to healthcare

Systems biology enables a comprehensive approach to studying biological systems (Ideker et al., 2001a; Huang and Hood, 2019), far different from the earlier reductionistic studies of biology one gene or one protein at a time. Systems thinking has emerged as a consequence of the complexity of biology and disease—and the possibility of deconvoluting this complexity by generating and computationally analysing appropriate types of 'big data'. Strategies to guide discovery from high-throughput data include machine learning/artificial intelligence (Miotto et al., 2018), and reconceptualization of fundamental ideas and mechanisms.

Systems thinking requires holistic attention across several domains of research and healthcare, including all biological systems within an individual as well as the environmental and sociological contexts that exert pressure on those systems. Systems thinking is:

1. Integrative and high dimensional—biological networks are integrated and hierarchal, operating at the levels of molecules, cells, organs, individuals, society, and interaction with the environment. The integration of different data types reveals interrelationships among these hierarchies, and can help separate signal from noise (Ideker et al., 2011). Each data type should be measured as completely as possible. This is relatively simple for the genome and transcriptome of specific tissues, since nucleic acid sequencing technology is adept at an unbiased cataloguing of all DNAs or RNAs present. However, other 'omes' (e.g. proteome, metabolome, physiome, envirome), while deeply informative, can be more challenging with regard to capturing all the data features since they typically employ targeted strategies that require defining in advance what will be catalogued. 'Big data' and machine learning can be employed to help make sense of the data and allow the derivation of statistical inferences about biological datasets. These inferences then lead to hypotheses about biology and disease that can be tested by systematic perturbation experiments on living cells, tissues, and individuals, as well as by computational simulations *in silico*.

2. Temporally dynamic—how systems change with time to reflect wellness and disease provides valuable insights into the mechanisms underlying these phenomena. The key is deep longitudinal phenotyping of individuals which means making thousands of measurements on blood analytes, the gut microbiome, and digital health measurements—say

every six months—for each individual. Making extensive longitudinal measurements to deeply phenotyped changes in hundreds of biological networks provides a '360 degree view' of individual health. Digital health (Evangelista et al., 2019) also provides increasingly useful devices to make frequent or continuous quantified self-measurements on features of human physiology (and disease). Continuous digital monitoring coupled with deep 'omic' phenotyping and whole genome analyses on individuals, with appropriate computational analyses identifies ultimately thousands of touch points for actionable possibilities that, if acted upon, can transform patients' lives.

3. Comparative—big data can be used to infer the networks (biological wiring diagrams) that mediate biology and disease (Hwang et al., 2009; Ideker et al., 2011). A comparison of normal and disease-perturbed networks enables identification of candidate biomarkers and potential drug targets, and the understanding of biological mechanisms of wellness and disease. Importantly, these analyses can also reveal individualized trajectories that show a diversity of biological paths from wellness to disease, even when symptoms are similar.

Many view wellness with scepticism because much of what is sold under the umbrella of wellness today lacks a scientific basis. Some of this scepticism is certainly justified, given that we spend the vast majority of our healthcare dollars (and a comparable percentage of our research dollars) on disease and very little on wellness. Thus, comparatively little rigorous scientific research has been carried out on the science of wellness. As Dennis Ausiello, Emeritus Chief of Medicine at Harvard Medical School's Massachusetts General Hospital has often said (quoted here with permission), 'Healthcare is the only industry that does not study its own gold standard, which is wellness.'

An evaluation of the determinants of human health over a lifetime are striking (Schroeder, 2007): lifestyle/environment accounts for 60%; genetics for 30% and state-of-the-art healthcare for just 10% (Papanicolas et al., 2018). A strong case can be made that 20th century medicine has been ineffective for 90% of health drivers, explaining the large costs without commensurate benefits (Milstein et al., 2011). Longitudinal deep phenotyping (phenomics) can assess the multimodal contributions of genetics and lifestyle/environment to health that are not well covered by our existing healthcare systems. Any deep understanding of wellness and complex diseases must deal with a far broader spectrum of the biological complexity of humans such as sex differences, differential environmental challenges, healthcare disparities and the like. Moreover, a major challenge is the lack of general metrics for gauging wellness and assessing its improvement or the earliest transitions to disease—for which we propose scientific wellness as a solution.

Scientific wellness

We define scientific wellness as the quantitative data-driven assessment of wellness aimed at optimizing, for each individual, wellness trajectories and avoiding or reversing disease trajectories. The first major need for developing scientific wellness is to generate large amounts of deep phenotyping data beginning with healthy individuals, following them longitudinally to understand and optimize wellness, and identify any disease transitions to learn how disease manifests in its earliest stages (Zubair et al., 2019).

Perhaps the best existing metrics for wellness are the blood biochemical markers that the medical literature has validated as indicators for specific diseases, and these often lead to actionable possibilities for improving health (e.g. haemoglobin A1c, glucose, low-density lipoprotein cholesterol, high blood lead, mercury, iron, toxins, etc.). However, individuals are biologically complex and reflect the integration of numerous environmental interactions with biological systems, and contemporary medicine assays very little of this complexity. With our standard approach to assay approximately 1200 blood analytes at each blood-draw time-point (these are the number of measurements we can obtain for a reasonable cost per individual and they have been more than enough to validate scientific wellness and the power of longitudinal data clouds) and the complete genome sequence, we have the ability to assess the operations of hundreds of different biological systems; and with appropriate longitudinal data to turn interesting observations (and hypotheses) into actionable possibilities for improving the health trajectories of patients.

Generating the data needed for scientific wellness: from $n = 1$ to $n = 1$ million

To achieve the goals of scientific wellness to enhance health and to predict and prevent disease, we need extensive data sets that are personal (a data cloud for each person because everyone is different), dense (lots of measurements to cover a wide range of biological systems), and dynamic (identify informative changes over time)—in other words longitudinal deep phenotyping. Deep phenotyping projects are on the rise, with orders-of-magnitude increases in their scale. The first deep phenotyping projects were centred on single individuals (n = 1) and came initially from Michael Snyder in relation to type 2 diabetes (Chen et al., 2012), Larry Smarr in relation to inflammatory bowel disease (Smarr, 2012), and Eric Alm in relation to the microbiome (David et al., 2014).

In 2014, Hood and Price initiated the Pioneer wellness 100 project (N~100; Price et al., 2017) to launch scientific wellness, where we coupled genome sequencing and deep phenotyping for discovery of actionable possibilities with health coaching of individuals (Hood et al., 2015; Zubair et al., 2019). This means persuading individuals to act upon relevant known actionable possibilities. We carried out a nine-month, IRB approved, pilot programme employing complete genome sequencing and deep phenotyping (on 108 individuals (Price et al., 2017) we called wellness pioneers) ranging in age from 21 to >89 years and relatively equally divided between males and females. This broad age group allowed us to draw fascinating conclusions about the biological age metrics (see below) and assess some interesting differences between young and old individuals. The resultant longitudinal data clouds enabled us to assess, for each individual, major determinants of health including genetics and lifestyle/environment. This study had several distinct features.

1. The deep phenotyping assays included a complete genome sequence, quantification of blood analytes every three months (about 1200 clinical chemistries, proteins, and metabolites), compositional analysis of the gut microbiome, digital health measurements (e.g. Fitbit), self-reported health data from questionnaires, and an assessment of environmental toxins (e.g. mercury, lead) detectable in the blood.

2. These efforts generated longitudinal data clouds for each individual that were analysed to identify actionable possibilities to improve wellness or avoid or ameliorate disease. These recommendations were culled from the scientific literature based on previous clinical trials and pointed to by the analyses of the individual data clouds. As the data clouds increased in number and were lengthened, many new actionable possibilities emerged, including those dealing with nutrition, inflammation, prediabetes, heavy metal toxicities, and often informed by the integration of multiple data modalities.

3. Wellness coaches, trained primarily in behavioural psychology and nutrition and working with the study physician, brought these actionable possibilities to each individual monthly and they were key to helping individuals carry out their actionable possibilities and attain their health objectives. Physicians also intervened when observations warranted medical attention—and these were referred back into the healthcare system. Another key point is that this process had the pioneers actively participating in optimizing their own health—the essence of the fourth P: participatory (Hood and Auffray, 2013).

The story of one participant in the study is illustrative. He lost 25 pounds through diet and exercise (and now is back at his college football weight). He was prediabetic and through a modified diet converted back to a normal blood glucose. He had extremely low vitamin D blood levels, as discussed 'Actionable possibilities', and brought the vitamin to a normal blood level. In a similar manner, as the study population increased, many new actionable possibilities were identified. And these health improvements across the study population were strikingly reflected in blood analyte improvements (Price et al., 2017).

Overall, the pioneers experienced significant improvements in their health (in blood analytes, weight loss, benefit from exercise, examples where disease was avoided or reversed, and a personal feeling of wellness). The vast majority wanted to continue beyond the nine-month programme. In part to enable this continuation, Hood and Price co-founded the company, Arivale, to bring scientific wellness to consumers—and many of the pioneers joined to continue their journey of scientific wellness over the next four years. Arivale employed the same strategies as described for the pioneer hundred programme, except that the blood draws and stool samples were taken every six months. Two notable accomplishments were (i) the recruitment of more than 5,000 clients and (ii) that their data clouds revealed more than 150 transitions from wellness to different diagnosed chronic diseases over the period 2015–2019 (for example, thirty-five of these were cancer transitions). Arivale ended in June of 2019 for two main reasons. First, it did not have an effective education programme to recruit new clients to scientific wellness, which is a new concept to most people. Thus, a population of clients large enough to make scientific wellness profitable was not attained—we estimate that we would have needed more than 20,000 clients to succeed. We have run a clinical trial on 1,000 patients with Providence. These data are currently being analysed and look promising in terms of striking improvements to blood analyte levels. The hope is that payers will recognize the power of scientific wellness to save them healthcare dollars and that they will routinely pay for scientific wellness in the future, thus removing one of its major constraints: cost. Second, the US Food and Drug Administration (FDA) dictated that genetics-to-consumer companies such as Arivale could not bring disease-related actionable possibilities to their clients (the FDA shut down 23andme for almost a year over a similar issue). This significantly limited the information that Arivale could act on that could have benefited their clients. Regardless, over the short life of the company, Arivale generated 5,000 longitudinal data clouds up to five years in length that provided several striking new insights into human wellness and disease, discussed below.

Actionable possibilities

With Arivale, as with the P100 project, the number of actionable possibilities increased significantly with the increasing number of data clouds and with the integration of different data types. For example, the participant in the P100 programme who had very low blood vitamin D, known to contribute to a plethora of conditions (Charoenngam and Holick, 2020), did not manage to change his low levels despite taking One thousand international units (IU) of vitamin D daily. We identified from the literature three genes, each with two variants that blocked the uptake of vitamin D. If an individual had two or more of these variants, large doses of vitamin D (10,000–15,000 IU) were required to move to a normal baseline and relatively large doses (5,000 IU) were required for maintenance. Thus, the integration of genome variants with blood clinical chemistry generated a new actionable possibility. Low vitamin D is associated with osteoporosis, Alzheimer's disease, cardiovascular disease and many others, as well as an increased susceptibility to some viral infections, including severe COVID-19) (Liu et al., 2021). Most physicians never check vitamin D blood levels.[1] This example suggests that if individuals were to take a life-long journey of scientific wellness, responding to all of the old and new relevant actionable possibilities as they arise, then there could be a chance to improve individual health trajectories so as to make them the equivalent to Eric Topol's 'wellderly'—individuals in their 80s

---

[1] On the controversy regarding the need to screen for vitamin D levels and its meaning, see Offit 2020, pp. 52–57. —The editors.

and 90s with no chronic disease and not taking medications. Scientific wellness thus provides an effective path to healthy ageing. We discuss below other ways to implement healthy ageing under this paradigm.

Biologically relevant statistical correlations

For the P100 pilot, we calculated approximately 3,500 statistical correlations between data types (analytes, genes, proteins, etc.; Price et al., 2017). This hair ball of statistical correlations could be broken down into more than seventy discrete communities of blood analytes and other data types. Each community correlated with a physiologically relevant feature or disease-related phenotype. For example, the cholesterol community had fifteen analytes and genetic markers—and some of these are wonderful candidates as blood biomarkers or even candidate drug targets for lowering blood LDL. This effectively provided insights into biological system connectivity not previously appreciated and generated new hypotheses for intervention touch points. Thus, one can use a statistical approach with deep phenotyping data to generate a new approach to biomarker and drug target discovery. This represents a striking new opportunity for Pharma to discover biomarkers and drug targets.

Genetic risk

Genome-wide association studies have generated polygenic risk scores (PRS) for more than hundred different diseases and disease phenotypes. These can be translated to individuals whose whole genome sequence has been determined to identify their relative genetic risks for these diseases. These polygenic risk scores are important in two ways. First, well people at high risk for serious diseases may be followed with longitudinal deep phenotyping to identify the earliest disease transition point, so that there can be an attempt to reverse the disease before it ever manifests itself and emerges as clinical symptoms. Second, Arivale was able to demonstrate that individuals at high genetic risk for elevated LDL cholesterol and who have high cholesterol blood levels were unable to bring their LDL down with lifestyle changes alone, suggesting that they likely need to use statins or other cholesterol-lowering drugs. Conversely, for individuals with a low genetic risk for elevated LDL cholesterol but high blood cholesterol regardless, lifestyle changes worked effectively to bring down the LDL level (Zubair et al., 2019). This is a beautiful example of how the one-treatment approach employed by 20th century medicine does not fit all patients. We contend that there will be many other traits for which individuals may be treated differentially, depending on whether they have a high or low genetic risk for a given disease.

Manifestation of analytes in the blood according to genetic risk

An additional use of PRS is to identify the blood analytes and proteins that associate with them. Since PRS can be calculated for all individuals and everyone falls somewhere on the risk continuum (even if that risk is extremely low), every person is relevant to the study of every disease for which PRS are available. We determined blood analyte–PRS correlations from the deep phenotyping data to identify analytes that associate with the risk of disease (as opposed to the emergence of disease) for fifty-four different polygenic scores. Changes in some of the associated analytes were also present in those who had developed the disease, suggesting that they represent biologically relevant physiological dysfunctions and that closer examination of those systems could be useful in predicting early transitions to disease (Price et al., 2017). This is important for both the identification of novel biomarkers of disease and to gain insight into early wellness-to-disease transitions.

The diversity of the gut microbiome is reflected in the blood metabolome

Increasing diversity of the gut microbiome reflects health for the individual, with higher species diversity (to a point) generally associated with better health. The Arivale data led to the discovery of a panel of blood metabolites that can predict the diversity of the gut microbiome (Wilmanski et al., 2019). Thus, the assessment of individual gut health may be possible by the simple quantification of these metabolites from blood, which is easier and less invasive to sample. The gut microbiome has already been implicated in many disease states (Zubair et al., 2019; Wainberg et al., 2020) and we are confident that there will be a continued expansion of these data. Thus, understanding how blood analytes can predict microbiome diversity could be another powerful tool for identifying early disease transitions.

Biological age

The Arivale population age range extended from 20s to 90s. We were able to demonstrate that the variance and levels of the expression patterns of the three classes of blood analytes (clinical chemistries, proteins, and metabolites) changed with age. From this observation, we developed an algorithm to calculate an individual's biological age (Earls et al., 2019)—the age your body says you are rather than the age your birthday says you are. If your biological age is younger than your chronological age, this suggests that you are ageing in a healthy manner. Importantly, while the slope of chronological age is locked into a linear increase, the slope of biological age can be flattened or even pushed into a decline. So, the difference between biological and chronological age is modifiable. For example, the 300 Arivale individuals with type 2 diabetes had, on average, biological ages six years greater than their chronological ages. Indeed, we looked at Arivale customers with forty different diseases and in each case their biological ages were greater than their chronological ages. Conversely, the individuals that fell into the upper 5% of physical activity (as determined by Fitbit data) had biological ages three years younger than their chronological ages. Individuals who came into the Arivale programme with biological ages of ≥5 years greater than their chronological ages lost, on average, one year in their biological age for every year they stayed in the wellness programme, thus reducing the difference between biological and chronological age, indicating the health value of the programme. Hence biological age may be a useful metric for assessing healthy ageing in a highly integrative manner. Moreover, the metabolite measurements we use to generate biological age provide insights as too how individuals may age more effectively with corrections of vitamin or supplement levels. Hood and Price have recently co-founded a company (Aevum Aging that has now been acquired by Onegevity) that will determine an individual's biological age from a blood sample and then provide insights as to how the healthy ageing can be optimized. We believe this will be a powerful approach to facilitate healthy ageing and thus help deal with the current ageing US population.

Wellness-to-disease transitions

Arivale had more than 150 individuals with wellness-to-disease transitions for many different chronic diseases. Thirty-five of these were transitions to cancer. As we examine the blood analytes for the transitions to different cancers, we can identify individual blood analytes (or their cognate disease-perturbed biological networks) that signal the earliest detectable onset of disease. With enough data and artificial intelligence to deal with significant signal-to-noise issues, we will be able to diagnose common chronic diseases well before they ever manifest themselves as a disease phenotype. In time, we will be able to employ a systems approach to understanding the biology of these early disease transition and create therapies that will reverse them at their earliest stages—and, in turn, this will be a component of '21st century disease prevention'.

A model for wellness and disease transitions

These wellness-to-disease transitions lead to a new way of thinking about wellness and disease. Longitudinal deep phenotyping raises the possibility that in time we could establish a baseline wellness trajectory for each individual in a given population: then detect the earliest wellness-to-disease transitions for most chronic diseases and reverse them before they ever manifest as disease phenotypes. We would view all disease as arising from this continuum from wellness to disease. Importantly, since this approach allows us to identify changes in trajectories of blood metrics, we can identify wellness-to-disease transitions even though they may look very different across individuals moving toward the same disease diagnosis—this is a very different concept of biomarkers, that can be highly individualized. We have demonstrated in a mouse model of neurodegeneration that, as the disease progresses from its earliest transition point, the number of disease-perturbed biological networks increases exponentially (Zubair et al., 2019). Accordingly, the focus of study for all chronic diseases should be at these earliest transition points—and not as it is done today—after the disease phenotype has emerged. As the number of disease-perturbed networks increases, disease becomes increasingly difficult to stop or reverse. Clearly, interventions that might be used in presymptomatic cases must be highly safe and precise to make them beneficial and cost-effective.

$n = 1$ experiments

$n = 1$ experiments use genome/deep phenotyping to follow the wellness and disease trajectories in a single individual. In a sense, the Arivale cohort constitutes 5,000 $n = 1$ experiments (Wainberg et al., 2020). There are many applications for these $n = 1$ experiments to:

(a) optimize wellness for each individual;

(b) detect wellness-to-disease transitions early and initiate reversal (we had several examples of cancer in this regard);

(c) understand complex human systems such as response to nutrition (every Arivale client optimized their nutrition uniquely—and hence we have 5000 $n = 1$ experiments for nutrition);

(d) follow the treatment of a complex disease (we did this for one bladder cancer example and it highlighted what an effective future therapy should be, as well as a failed current therapy);

(e) use longitudinal deep phenotyping for clinical trials. Indeed, we are currently carrying out such clinical trials on multiple sclerosis (with Genentech), scientific wellness, cancers (ovarian, colon, and breast), Alzheimer disease, and COVID-19. We also have an ongoing clinical trial with breast cancer survivors that will look at how to ameliorate the many side-effects of chemotherapy and irradiation.

The coming tidal wave of data and the PSJH million person genome/phenome project

Additional deep phenotyping efforts on small populations by other groups have appeared in recent months with $N \approx 100$, most notably high-profile efforts led by Michael Snyder (Schussler-Fiorenza Rose et al., 2019; Zhou et al., 2019). There have also been comparably sized unpublished studies carried out now in Finland and Sweden. There are also major international efforts, most notably in China, that have committed major resources to scale their national deep phenotyping programme, establishing the International Phenomics Project, co-led by Felix Li (China), Lee Hood (USA), and Jeremy Nicholson (Australia).

Several programmes are aimed at genomics analyses now at $n = 500,000$ to 1 million, although with deep phenotyping not yet implemented. The NIH All of Us Program seeks to generate genomic data for 1 million people in the USA and link these genomes with medical, health information (especially electronic medical records and data from digital wearables) and socio-economic factors. By banking blood, this becomes a community resource that over time can become deeply phenotyped as groups perform different large-scale data generation from the same samples. A similar large-scale and highly influential cohort has been generated on 500,000 people in the UK Biobank (Sudlow et al., 2015) that again provides genomic analyses with banked samples that can be later phenotyped in a multitude of ways. Such large-scale biobanks with shared data can become examples of deep phenotyping as more and more longitudinal measures are done across the same individuals and these can be analysed together in integrated analyses.

PSJH million person genome/phenome project

Spearheaded by ISB and PSJH, which serves 10 million people across fifty-one hospitals, we are initiating a programme for a planned million person genome/phenome project over five years while biobanking samples to scale the scientific wellness, healthy ageing, and deep phenotyping approaches for both well and sick patients. In 2020, we began the first stage of this project by initiating whole genome sequencing and deep phenotyping of 5,000 individuals who are suffering from common chronic diseases (~4,000) or who are well (~1,000). This pilot project will help to define the logistical, clinical and research questions for the larger project at scale. The increase of actionable possibilities has grown exponentially through discovery science from the Pioneer 100 and Arivale cohorts. We are confident the same increases will continue to hold true as we scale towards $n = 1$ million. However, generating these data is only the first step, as myriad challenges remain in re-orienting healthcare systems more towards wellness and the prevention and early-stage reversal of disease processes.

Technology development

Striking advances in measuring genomics, metabolomics, proteomics, the microbiome, physiological features, and digital assessments, all in individuals, are increasing the data throughput and decreasing strikingly the assay costs. For example, the single-molecule nanopore analyses of Oxford Nanopore (and the equivalent from other companies) offer the opportunity for long reads and assembly of genomes independent of highly inaccurate reference genomes. Such an approach will provide accurate HLA and pharmacogenomic data, read the methyl modified Cs from unamplified

DNA[2], and will allow direct sequencing of full length mRNAs—greatly expanding the depth of information from the genomics piece of deep phenotyping. Aptamers are being used to identify and quantify thousands of proteins including for wellness and disease-monitoring applications (Sun et al., 2018; Williams et al., 2019), as are DNA-based assay read-outs of protein enzyme-linked immunosorbent assays (Price et al., 2017; Wilmanski et al., 2019). There is also the intriguing possibility of generating DNA-based readout assays for some metabolites (Augspurger et al., 2018). Such DNA-based assays will increase the assay throughput and substantially reduce their cost—with the potential of bringing them into a single platform. An exponentially increasing array of digital devices is becoming available, enabling many different types of physiological measurements. Also critical are the algorithms and computation platforms necessary to analyse, integrate, and derive actionable possibilities from diverse types of data. Blood is being viewed as a window into health and disease, and it is increasingly possible to measure globally various analytes (clinical chemistries, proteins, metabolites, lipids, liquid biopsies, etc.) as well as to perform single-cell analyses and assessment of the contents of exosomes shed from many different organs and cell-free DNA. The key developments necessary for the wide-scale adoption of deep phenotyping in clinical settings is increased throughput, increased global (comprehensive) measurements for proteomes, metabolomes and microbiomes, and substantially decreased costs.

Challenges to turn research into clinically actionable possibilities

The challenges of implementing scientific wellness into 21st century medicine are many, and include:

- the need for a much lower cost of genome and phenome assays;
- making the assays for proteins, metabolites, and various small molecules (e.g. lipids) as comprehensive as possible;
- creating effective training programmes for scientific wellness coaches and physicians whose abilities will be amplified by apps and avatars;

---

[2] An epigenetic marker. —The editors.

- employing the powerful tools of machine learning to assess patterns in the data clouds and artificial intelligence to create learning systems that can readily convert data into actionable possibilities;
- generating new and inexpensive assays to explore in far more detail many new dimensions of human physiology and disease, e.g. the brain, the immune systems, integrated hormonal systems, etc.

A key factor to consider that arises with the measurement of multiple analytes is that of multiple hypothesis testing and false positives (Omenn et al., 2012; Sung et al., 2012). These issues can be dealt with increasingly effectively as the population size increases. More importantly, longitudinal deep phenotyping enables different strategies for error correction of potential false positives, because a single outlier in a biomarker can be judged by whether or not there are other corroborating changes seen across the data, where this cannot be done in the absence of other data. For this reason, we would argue that high sensitivity and specificity in monitoring for a wide swath of diseases is only ultimately possible with deep phenotyping of individuals. Another critical point to keep in mind is that the necessary safety threshold of an intervention goes up the earlier it is deployed. For this reason, comparisons to screening for cancers where the common remedy is toxic chemotherapy would not be a good comparison for scientific wellness, because such a damaging therapy would never be deployed in a preventive manner or in a regime with high false positives. We also believe that scientific wellness should be extended to infants and children during their most formative years for improving their lifelong trajectory of wellness (Barker et al., 1993). Scientific wellness should also include brain health as well as body health (Nahum et al., 2013; Merzenich et al., 2014). Alzheimer's disease and related dementias are one intriguing example where we know that wellness approaches rooted in an appropriate lifestyle can have a significant effect on reducing incidence (Ngandu et al., 2015), while we unfortunately still lack an effective drug treatment for later stage disease.

Challenges to implementing scientific wellness in the healthcare system

As with any medical paradigm change, the healthcare establishment is generally conservative and, understandably, highly reluctant to accept and implement changes, leaving several challenges to surmount.

(1) Enlightened leadership (organizationally and operationally). Most healthcare leaders are primarily driven by optimizing profit centres and minimizing cost and not by innovation or pioneering medicine of the future. They generally do not realize how important it is to invest in the research, both to improve the quality of medicine and play a leadership role in inventing a future of medicine that will both improve healthcare and reduce its cost. What is needed are leaders who understand the 21st century medicine vision, embrace it, and accept its changes.

(2) The intrinsic inertia of the 3.8 trillion-dollar healthcare industry. It is highly regulated and very much driven by both private and government payer dictates. A single-payer system could have enormous advantages in aligning health and financial incentives as do integrated provider–payer healthcare systems. It is a compelling argument for large healthcare systems to become integrated payer/provider organizations as the integrated entity will accrue the striking savings of scientific wellness (these cost savings generally will go to payers). We are now analysing a scientific wellness clinical trial of almost 1,000 patients and believe from these compelling results will gradually come payer support for wellness.

(3) The need to educate healthcare professionals, patients, regulators and payers as to the nature of this new wellness-centric 21st century medicine is paramount.

Measuring the effectiveness of scientific wellness

So, how will we know that a scientific wellness-centric approach has succeeded or not? The scientific wellness clinical trial has been mentioned, and this is where assessment must start. There will be the usual trade-offs between 'hard' clinical endpoints that take time to verify and 'soft' endpoints, such as improvement in blood clinical markers, that are relatively quick to see but don't always translate to improvements in the 'hard' outcomes. Thus far, people who went through the Arivale scientific wellness programme had marked improvements in clinical labs (Zubair et al., 2019) and in estimates of biological ageing from multi-omics data (Earls et al., 2019). Importantly, the effect size of lifestyle intervention on changing clinical markers was strongly affected by genetics (Zubair et al., 2019), supporting the need to integrate across data types.

Personalized interventions based on data clouds can also be studied with randomized controlled trials. We have trials (some RCT and some observational) ongoing now using deep phenotyping, coaching, and medical interventions with clinical groups now focused on dementia/Alzheimer disease, breast cancer survivorship, Lyme disease, and multiple sclerosis, the latter funded by Genentech. The key metrics that will ultimately determine the success of the scientific wellness approach will be on measuring improved outcomes and total cost in the long run. Additionally, for a particular individual, whereas causality cannot be definitively assessed, deep data can provide a wealth of information related to that person's trajectory. When coupled with mechanistic understanding of, for example, a disease or wellness process, it is possible to gain significant insight into the processes that are shaping an individual's biology to a degree that is not possible with sparse data or a simple biomarker (Huang and Hood, 2019).

Defining the future

We predict that 21st century medicine and scientific wellness/healthy ageing will greatly improve the healthcare system. It seems logical to conclude that being well will reduce the cost of healthcare because it permits one to optimize health and avoid disease. However, this is just the start of the potential benefits engendered by scientific wellness. Many of these benefits have been discussed, including: discovering actionable possibilities through the integration of multi-omic data; leveraging polygenic risk scores to predict disease and discover new therapeutic avenues; predicting the relationships of the gut microbiome to health; using biological age as an indicator of health and success of intervention programmes designed to improve wellness, as well as optimizing healthy ageing; and identifying and reversing the earliest wellness to disease transitions, which may be unique across individuals. These have given us a solid base to launch into the future, which is already being realized.

21st century healthcare

With the background of this chapter we can now define the essence of 21st century medicine. It should assess the health trajectories of everyone separately (Figure 3.1). The idea is that each individual has three distinct stages to their health trajectory. First, most of us start with a wellness trajectory and a major role of 21st century medicine will be to optimize wellness. Second, as we age, the probability of transition from wellness to disease increases. The objective of 21st century medicine will be to identify these transitions at their earliest detectable stage, to identify biomarkers for this transition, and then to use a systems approach to identify the earliest disease-perturbed networks so that therapies can be designed to reverse disease before it ever manifests as a disease phenotype. Third, there will be some individuals who transition into disease and 21st century medicine should use systems approaches to design effective therapies—with an emphasis on treating just as soon as possible after the initial wellness-to-disease transition. This is an $n = 1$ medicine where each individual is treated as an individual (and not by the rules of population averages). By contrast, contemporary or 20th century medicine focuses on the disease state after a clinical diagnosis has been made—long after the disease was initiated and at a stage where there are many disease-perturbed networks—making simple therapies difficult, if not impossible (Figure 3.1).

Let us now consider one example of a chronic disease that we are attempting to treat using a 21st century medicine approach.

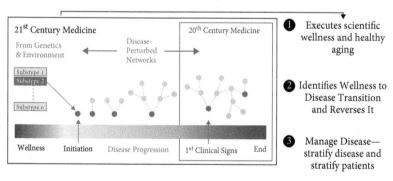

Figure 3.1 Twenty-first century medicine (healthcare). Follow the trajectory of each individual which divides into three phases: wellness, transition to disease, and disease progression. The preventive medicine of the 21st century will identify the earliest disease transitions and reverse them before they ever manifest as a disease phenotype. Contemporary or 20th century medicine focuses primarily on disease after clinical diagnosis—at a late stage in its progress. Twenty-first century medicine will focus primarily on optimizing wellness and health ageing—but will deal with transition and disease, employing a systems approach where necessary.

Alzheimer disease and brain health—an initial case in dealing with early reversal of a chronic disease

Major previous studies have reported positive effects of a healthy lifestyle (FINGER; Ngandu et al., 2015) and cognitive training (ACTIVE, ADDIN EN.CITE; Rebok et al., 2014) on delaying or improving cognitive decline in older adults. As with every complex disease, Alzheimer disease (AD) almost certainly has different biological trajectories that underpin the disease, even though clinical endpoints may appear quite similar. There are many genes (perhaps dozens to hundreds) that contribute to the risk and progression of AD, as well as a plethora of environmental factors (viruses, molds, chemical toxins), behaviours (diet, exercise, sleep, cognitive challenges) and resultant vitals and morphometrics (body mass index, blood pressure, glucose control). Thus, to expect that a single drug will effectively treat this (or any) complex disease seems naïve—and in fact approximately 400 AD drug trials in the last twelve years for single drugs have been attempted and all have failed. We are using the principles of P4 healthcare and scientific wellness in several prospective clinical trials to test the efficacy of multimodal interventions (diet, supplements, physical activity, cognitive training, chelation therapy, pharmacotherapy) on reversing cognitive decline in adults with mild cognitive impairment. Importantly, each patient's therapeutic protocol is based on their own deep phenotyping assessment of what they are missing—with some components being currently clinically actionable and others being developed by our research platforms to better understand the individualized disease trajectory and discover new approaches to treat it. Independently, we are collaborating with others to design and build computational models that will allow in-silico experimentation on individuals by inputting personal data clouds to predict outcomes of given interventions.

One can envision the identification of individuals at high risk for AD (e.g. individuals homozygous for the ApoE4 allele); following them with deep phenotyping to identify blood biomarkers for the earliest transitions and then employing the multimodal therapy for disease reversal.

Integrating everything with the million person genome/ phenome programme will enable 21st century medicine

The current clinical trials employ thousands to tens of thousands of experimental (disease) and control patients, appropriately randomized, to assess placebos and drugs. These trials generally cost in excess of one billion dollars. A fundamental flaw of the conventional approach is the unstated assumption in the designs that patients are identical—and, of course, this is not true as each patient differs from the others by their distinct genetics and lifestyle/ environments. The results of such a trial are aptly illustrated by an analysis of the ten top-selling drugs in the USA today. The responders to these drugs range at best from one in four to at worst one in twenty-five (Schork, 2015). That is, most of the patients waste their healthcare dollars and are exposed to the toxicity of the drugs without benefit. A P4 approach to clinical trials might be two-stage. First, around hundred patients are included in a trial employing genome sequencing and deep phenotyping. This will enable determination of blood biomarkers for distinguishing patients who respond to the drug from those who do not. Then a second trial of, say, hundred patients, all responders, could be initiated. Since the response rate to the drug would be close to 100%, the FDA should approve the drug for use with the companion biomarker diagnostic even with the small patient numbers (as it did for Genentech with Herceptin—a breast cancer drug (personal communication)).

Success of the million person project will establish a biobank of longitudinal samples that can serve as a reservoir to define nearly any cohort for retrospective clinical trials. These approaches could save close to one billion dollars for each approved drug. These savings could be carried across the drug development industry and to the consumer, as well as incentivize pharma companies to pursue small market drugs for rare diseases.

Finally, digital health is going to make many real-time, wellness-relevant health measurements at exponentially decreasing costs. As increasingly sophisticated health measurements are made we will have extremely informative continuous monitoring of physiological parameters. For example, heart rate variability—which integrates the activities of the sympathetic and parasympathetic nervous systems—can provide insights into both wellness and disease issues when coupled with deep phenotyping measurements. An additional possibility is to optimize the yields of the omic analyses—reducing their data dimensionality to perhaps 5,000 measurements to provide 99% of the information we need for scientific wellness. We envision that this will be measured at home with a 'tricorder' that could use a drop of blood or saliva. Then the data would go to a data centre for the determination of the actionable possibilities that in turn would be sent to the patient's physician or health coach and ultimately to the patients. We are just beginning to explore these possibilities.

Collectively these examples will lead to significant healthcare savings as the approaches are initiated in the healthcare system.

Coda

Twenty-first century medicine will be predictive, preventive, personalized, and participatory. The four Ps will be realized by practising genome analysis and longitudinal deep phenotyping (leading to the optimization of scientific wellness and healthy ageing) on a large population of presumably normal individuals as well as diseased individuals—the PSJH million person genome/ phenome project. This approach will allow us to address the four most challenging problems of 20th century medicine—quality of life, ageing population, chronic diseases, and cost. Twenty-first century medicine will move the healthcare systems to a value-based approach to healthcare. We need to educate healthcare leaders and physicians who are willing to adopt a 21st century healthcare agenda—both at the organizational and the operational levels.

Acknowledgements

We thank our colleagues at ISB for many stimulating conversations. We thank NIH, NSF, DOE and DOD for research support. Thanks to Sheryl Suchoknand for help on the manuscript and with life.

4

The problematic side of precision medicine

A short voyage through some questions

Giovanni Boniolo

Introduction

'Molecular medicine', 'precision medicine', and 'personalized medicine': do these expressions indicate something different? Even if one reads only some of the numerous papers using these terms, it is apparent that different research and clinical communities define the same *definiendum* (what is defined) with very slightly different *definientia* (by means of which we define) or use the same *definiens* to define different *defienenda*. For example, the Precision Medicine Initiative of former US President, Barack Obama (2015), calls precision medicine an 'innovative approach that takes into account individual differences in people's genes, environments, and lifestyles',[1] in order to deliver 'the right treatments, at the right time, every time to the right person', as Obama himself added in an interview (Kaiser, 2015). But these words and concepts are more or less the same we find in the *definiens* of 'personalized medicine'[2] and in the *definiens* of 'molecular medicine'.[3] In other words, the literature reflects the many different, albeit sometimes only slightly different, conceptualizations of personalized medicine and precision medicine. To cut the definitory 'Gordian knot', I will hereafter use the definition provided by the US National Research Council (NRC): 'According to the Precision Medicine Initiative, precision medicine is 'an emerging approach for disease treatment and prevention that takes into account individual variability in

---

[1] See also Ashley (2015), Collins and Varmus (2015), Kohane (2015).

[2] See, for example, the position paper of European Society for Predictive, Preventive and Personalised Medicine (EPMA) (Golubnitschaja et al., 2016).

[3] Molecular medicine is, however, somewhat different—a point that is not dealt with in this chapter (see Boniolo, 2017a).

Giovanni Boniolo, *The problematic side of precision medicine* In: *Can precision medicine be personal; Can personalized medicine be precise?*. Edited by: Yechiel Michael Barilan, Margherita Brusa, and Aaron Ciechanover, Oxford University Press. © Oxford University Press 2022. DOI: 10.1093/oso/9780198863465.003.0004

genes, environment, and lifestyle for each person.' This approach will allow doctors and researchers to predict more accurately which treatment and prevention strategies for a particular disease will work in which groups of people.[4] All of this will be enabled by new sequencing technology, molecular imaging technologies, and the major role of bioinformatics and the related computational tools.

Adopting the NRC definition of 'precision medicine', I present some of the main quandaries posed by this approach to biomedical research and clinical practice.[5] In 'The dark side of precision medicine', I briefly review: (i) how much information provided by PM is actually used; (ii) what we should know about PM; (iii) the patients-in-waiting question, and the corresponding medicalization of society. In 'Precision medicine and precise drugs', I discuss some ethical problems concerning PM and precise drugs.

The main aim of this chapter, however, is to present a brief selection of socio-ethical and epistemological problems that should be discussed at public level to help medicine remain loyal to its commitment to improve the quality of life of present and future patients.

The dark side of precision medicine

From information to knowledge

Let us begin with the emphasis that Barack Obama's inaugural speech placed on the idea that PM should increase our quality of life because it induces changes in our lifestyle. This is because a PM approach should provide comprehensive information about the relationships between genome, epigenome, the environment in which we live, and our lifestyles.

However, more information does not automatically increase knowledge. We should recognize that knowledge is substantially different

---

[4] See 'Precision medicine; on the relations between precision and personalised medicine', https://ghr.nlm.nih.gov. https://ghr.nlm.nih.gov/primer/precisionmedicine/precisionvsperso nalized.

[5] Some of the ethical problems posed by Precision Medicine are reformulations of problems arising with the advent of molecular biology, genomics and post-genomics, and even other issues existing before that. For example, Precision Medicine could have to do with questions that nowadays could be considered as classical bioethical topics, such as consent, privacy, the right to know and duty to inform, reproductive choices, incidental findings, distributive justice and money allocation, genetic screenings, overdiagnosis and overtreatment. Some of these, as we will see, acquire new significance in Precision Medicine.

from information. Information is the set of well-assessed and meaningful data.[6] Knowledge, on the other hand, concerns both the ability to retrieve logically structured and relevant information, along with the capacity to manage and even modify, and augment it. In the medical context, knowledge is relevant when it can contribute to people's health and to the related quality of life. Thus, on the one hand, there is the information provided by Precision Medicine concerning, for example, our genomic profile; on the other, there is our capability to transform it into knowledge to increase our quality of life and the quality of life of our relatives, friends, fellow citizens, etc.

In light of this distinction, Obama's enthusiasm should perhaps be tempered because, as Caulfield (2015) notes, it is by no means a given that 'genetic risk information will cause us all to exercise more and stop eating too much'. *In primis*, even if the information concerning our risk factors could increase our appreciation of a healthier way of living (see Wiseman, 2010), we do not need PM to know that abstinence from smoking, eating healthy food, and doing exercise is good for health. Moreover, there is a strong body of research in sociology and psychology showing that if people have information about the fact that something is bad for their health, it does not necessarily imply that they will act accordingly (Cook and Bellis, 2001; Kelly and Barker, 2016). Thus, even in the case of Precision Medicine, more information does not necessarily entails actionable knowledge.

Furthermore, there is something apparently paradoxical in relation to the practical value of information. Sometimes, the more information we have, the greater is our uncertainty about how to act on this information. This is perceived also at the layman's level, as many articles in newspapers and magazines show (Joyner, 2015; Welch and Burke, 2015). The enormous number of genetic and epigenetic variants discovered every day and whose phenotypic or pathologic significance is unclear, the spatial and temporal heterogeneity found in pathologies like cancer (Boniolo, 2017b), the clinical difficulty of interpreting the deluge of data produced, and the incapacity to master the many probabilistic causal paths due to mutations are examples of situations that should spur us to rethink the quality and usefulness of the mass of information available and how we should manage it to increase our knowledge. Not only are these issues relevant to medicine but also to wider humanistic reflections as the passage from information provided by Precision Medicine and the capacity to govern it are significant issues.

---

[6] For a detailed discussion of this point, see Floridi (2013).

The knowledge required by precision medicine

Numerous papers warn that many clinicians lack understanding of Precision Medicine and its molecular and biotechnological bases (Collins, 1997; Metcalfe et al., 2002;; Biesecker et al., 2012; Raghavan and Vassy, 2014).[7] This implies an urgent need to revise medical education, including postgraduate studies (Siddiqui, 2003; Teunissen and Dornan, 2008; Tabe, 2015; Dzau and Ginsburg, 2016; Friedman et al., 2016; Goldman, 2016).

Additionally, in the age of patient empowerment, self-diagnosis, fake-news and questionable sources, proper awareness of the promises and limitations of Precision Medicine is lacking (Rankin et al., 2005; Bastable et al., 2011; Hansberry et al., 2014, 2015). In particular, we should become aware of the distance between the announcement of a laboratory research discovery and its maturation at the level of good clinical practice.

There is another point to be considered. Patients' own choices are often swayed by learning about a *possible* onset or of a *possible* development of a disease. Possibility, i.e. probability, is therefore at the heart of much decision-making not only as regards diagnostic or therapeutic paths but even lifestyle plans. This implies that not only patients understand probabilistic information, but also that the clinicians who communicate this information to their patients have clear understanding thereof. Wegwarth et al. (2012) show lack of such understanding among American oncologists, in particular regarding the proper meaning of the cancer screening statistics.

Physicians have a duty to know the philosophical foundations on which their work rests. Patients should understand physicians' probabilistic communications because of the impact on their lives. This is highly pertinent when mathematical integration of diverse risk factors play central roles in preventive and curative medicine.

For example, let us consider molecular screening programs. They are laboratory procedures aimed at detecting specific molecular markers (e.g. the prostate-specific antigen, the mutated BRAC1, etc.) which indicate the probable presence or future development of diseases (e.g. prostate cancer, breast cancer, etc.).

Such markers may stand for deterministic or probabilistic chains of causation. In the first case, any time we have a causal event, we inevitably have the effect event (even if we do not know exactly when). For example, any time we

---

[7] See also https://www.nih.gov/news-events/news-releases/new-report-offers-primer-doctors-use-clinical-genome-exome-sequencing (accessed 30 April 2017).

have the mutation of the huntingtin gene (the causal event), we deterministically have Huntington's disease (the effect event). In other words, the probability of having Huntington's disease given the mutation of the huntingtin gene is equal to 1:

$$P(\text{Huntington's disease/mutation of the huningtin gene}) = 1.$$

Here the causal event is both necessary and sufficient.

In the case of probabilistic causality, any time we have a causal event, we increase the probability of having the correlated effect event. For example, the detection of an MYH7 mutation (the causal event) increases the probability of having hypertrophic cardiomyopathy (HCM) (the effect event). That is, the probability of having HCM given the MYH7 mutation is greater than the probability of having HCM in the case of a non-mutated MYH7:

$$P(\text{HCM/mutated MYH7}) > P(\text{HCM/non-mutated MYH7}).$$

In this case, the mutated MYH7 is necessary but not sufficient, reflecting incomplete penetrance. The same applies when we claim that HCM probabilistically causes sudden cardiac death (SCD). That is:

$$P(\text{SCD/HCM}) > P(\text{SCD/non-HCM}).$$

In this case, we have neither a necessary nor a sufficient condition. SCD may be caused by something different from HCM, for example, by coronary artery abnormalities, or by long QT syndrome.

Thus, not only should clinicians know the foundations of probability and statistics but they should also help potential and actual patients to comprehend the real statistical and probabilistic meaning of the outcomes they receive when they undergo a genetic or epigenetic test (Gigerenzer, 2007; Boniolo, Teira, 2016; Ferretti et al., 2017). In other words, there is a need to train them to comprehend what a probabilistic causal correlation is, how it differs from a deterministic causal correlation, and that a statistical correlation is not a causal correlation. These are surely crucial epistemological and clinical and aspects, but they have to do also with existential questions since they regard living choices (Boniolo and Campaner, 2019).

Implementation of mature initiatives of Precision Medicine requires not only knowledge concerning the molecular basis of pathologies, their treatments, and the biotechnology involved, as well as profound knowledge of the mathematical foundations of probability. No less crucial is caregivers' capacity to communicate these issues to the laity. The lack of such knowledge might not only be an epistemic weakness. It could also imply an ethical failure since the clinicians would not have the capacity to communicate genetic risks to ordinary patients.

Patients-in-waiting

Patients-in-waiting is the term used for the group of individuals under medical attention because they carry a risk of developing a disease. The patients-in-waiting category has come about because

> for some conditions no treatment is available; for others treatment is not as straightforward as dietary change but involves invasive medical procedures such as bone-marrow transplantation; still others constitute extremely rare and ill-understood conditions; for still others early detection has no discernible health benefits. Moreover, rather than diagnosing asymptomatic patients with clear-cut diseases, expanded [. . .] screening has identified [. . . individuals] with screening values outside a pre-set normal range that do not always clearly correlate with defined disease categories (Timmermans and Buchbinder, 2010).

The patients-in-waiting condition has been magnified by new sequencing and molecular imaging technologies, in other words, by the advent of Precision Medicine. Each us could be a patient-in-waiting if tested for a broad enough panel of conditions and profiles of risk. On average, every person carries one or two mutations that, if inherited from both parents, may cause severe genetic disorders or death before reaching reproductive age: 'Each haploid set of human autosomes carries on average 0.29 (95% credible interval: 0.10–0.84) recessive alleles that lead to complete sterility or death by reproductive age when homozygous' (Gao et al., 2015). In addition to inborn mutations, on average, each human acquires about 74 de-novo single nucleotide variants per genome (Veltman and Brunner, 2012).

Everybody is a potential (and sometimes even actual) patient-in-waiting. The chief implication is that 'patient-in-waiting' is a phenomenological reality (i.e. it is a real experience for the person) of a probabilistic title. Understanding this condition is understanding probabilistic knowledge. The ensuing implication is that if a condition "in waiting:" justifies actions (e.g.

medical interventions, personal preparation), this will expand the number of persons who need to act on that title, leading to the medicalization of society. This concern, which was raised almost fifty years ago (Illich 1976), and has now become even more pressing (see Conrad, 2007; Downing, 2011).

The warning concerning the possible medicalization of society runs parallel to the claim that there is "too much" medicine: 'This dramatic rise is best explained by increased testing and improved diagnostic tools, rather than a real change in cancer incidence. It has been described as an epidemic of diagnoses rather than a true epidemic. Similar 'epidemics' have occurred in conditions where there has been active screening, such as breast cancer and prostate cancer' (Glasziou et al., 2013).[8]

'Too much medicine', that is, overdiagnoses and overtreatments, or simply unnecessary treatments, could be the unwanted, undesirable but inevitable negative side-effect of Precision Medicine. Not only does all of this imply an increase in medical expenditure, but also an increase of anxiety and stress for individuals.

To paraphrase Max Weber from some decades ago (Weber, 1917), the more information we have, the more we should eat the 'fruit of the tree of knowledge'. Being a patient-in-waiting or having too much medicine could be the unwelcome outcome of having plenty of information about our genetic and epigenetic structure. They are the potential 'bitter fruits' of Precision Medicine that require great societal, ethical, and political attention (see, for example, Gefenas et al., 2011; Ormond and Cho, 2014; Fiore and Goodman, 2015; Blasimme and Vayena, 2016; Boniolo and Sanchini, 2016; Hammer, 2016; Nicol et al., 2016).

Precision medicine and precise drugs

We know that one of the main goals of PM is the production of new drugs addressing precise molecular targets. This has had beneficial outcomes, as in the case of certain cancer treatments. The discovery of the different hormone receptors and the diversity of their expression among breast cancer tumours have altered the diagnosis, treatment, and prognostication of breast cancer (Zardavas and Piccart-Gebhart, 2015).

---

[8] See also the BMJ initiative that 'aims to highlight the threat to human health posed by overdiagnosis and the waste of resources on unnecessary care': http://www.bmj.com/too-much-medicine (accessed 30 April 2017).

In the next section, I will show, there are two kinds of problem. The first is due to erroneous clinical trial design whose negative consequences are magnified by the PM approach. The second is strictly linked to the intrinsic features of PM, which could lead to injustice.

Clinical trial design and economic justice

In 2004 cetuximab (Erbitux) was approved as treatment to colorectal cancer patients whose tumour cells tested positive to the EGFR (epidermal growth factor receptor) receptor. Hey and Kesselheim (2016), who narrate this story, point out that the clinical trial that led to the approval did not enrol any EGFR-negative patients. There was, therefore, no control group! Later it was discovered that EGFR expression is irrelevant. Another marker was found predictive of response, that is, KRAS (Kirsten rat sarcoma viral oncogene). In the end, patients who would have benefited from the drug did not receive it, while others received it with no positive effects—and possibly with negative effects. Surely this is a case of poor clinical trial design, but it is also a warning towards the focus only on a particular biomarker.

In 2015, a research group proposed a randomized controlled trial—known as SHIVA[9]—to compare a PM approach based on particular biomarkers with conventional therapy. Possibly this was the first clinical trial testing Precision Medicine's contention of treatment selection by biomarkers (Le Tourneau et al., 2015a and 2015b). The trial was conducted on 195 patients with metastatic cancer (any solid tumour) refractory to standard care. Patients were randomly assigned to receive either molecularly targeted agents chosen on the basis of the molecular profile of the tumour, or therapy of the physician's choice. Median follow-up period was 11.3 months. Median progression-free survival was of 2.3 months for patients receiving targeted therapy versus 2.0 months for patients receiving therapy based on the physician's choice. The authors conclude: 'So far, no evidence from randomised clinical trial supports the use of molecularly targeted agents outside their indications on the basis of tumour molecular profiling'. The paper generated vehement controversy regarding its interpretation.[10]

---

[9] See https://clinicaltrials.gov/ct2/show/NCT01771458 (accessed 30 April 2017).
[10] This paper was heavily commented both negatively and positively (Tsimberidou and Kurzrock, 2015; Hahn and Martin, 2015; Weiss, 2015; Le Tourneau et al., 2015; Le Tourneau and Kurzrock, 2016).

Certainly, a negative result in specific trials does not exclude the possibility that precision-tailored treatment might work in other circumstances or as first-line therapy. Nevertheless, it is fair to conclude that Precision Medicine's approach does not fit many medical conditions and treatment dilemmas, and that determination of treatment by molecular markers is not prima facie superior to prevailing criteria of treatment recommendations. Moreover, greater attention should be given to the design of clinical trials in the age of Precision Medicine (see, for example, Saad et al., 2017).

This last step has already been taken and it has brought to light several different kinds of new experimental designs (or has led to the rethinking of old ones), such as longitudinal cohort studies with or without downstream clinical trials, studies assessing the clinical utility of molecular profiling, master-protocol trials, basket trials, adaptive trials, N-of-1 trials, window-of-opportunity trials, etc. (Heckman-Stoddard and Smith, 2014; Zardavas and Piccart-Gebhart, 2015). This means a lot of work—largely still to be done—for the ethicists and philosophers of science, who can apply their expertise to the analysis of the ethical plausibility and the epistemological and methodological soundness of these new approaches to drug validation (Teira, 2017).

Personalized Medicine raises issues of justice too. Lumacaftor/ivacaftor is a drug combination approved by the US Food and Drug Administration to correct the defect of preventing the regular passage of salt molecules through cell membranes. This combination was studied on, and consequently approved for, a small fraction (about the 5%) of patients with cystic fibrosis who are identifiable by genetic testing. But we should also take a look at the other side of the moon. If only 5% of patients with cystic fibrosis can benefit from it, the other 95% do not. Moreover, drug approval lasted decades, and treatment costs were $300,000 per year per patient. Besides, a study published in the *New England Journal of Medicine* found that the extent to which lumacaftor/ivacaftor helps its target patients is more or less equal to that of the conventional and one-size-fits-all treatment (a combination of ibuprofen, aerosolized saline and azithromycin), whose cost is $300 per year (Rehman, Baloch, et al., 2016; see also Interlandi, 2016). This finding prompts a series of questions. Is it really necessary to invest time and money in a specific drug whose effect is comparable to a classical drug? Is it really fair to invest time and money in a given small percentage of the patient population knowing that the majority of patients will consequently not have an effective drug? On the other hand, is it really fair not to invest money in a given small percentage only because it is small? Who chooses to allocate money to a tiny fraction or to the majority of patients? In particular, should we take into consideration—whenever we plan a clinical trial in the age of Precision Medicine —the number of potential patients who could benefit from the new drug?

The 'race' question and the unknown genotypes

Whereas targeting 'heart failure' or cystic fibrosis for research and clinical benefit is an evident social good, the choice of genes or molecules is much less obvious. Is the EGFR receptor worth millions of euros of investigation, or perhaps the KRAS? In the absence of evident common goals such as the battle with heart failure, the minute division of research and care (the process of rendering medicine more 'precise') is liable to bias and confusion. A case in point is the attempt to slice 'heart failure' by race, a much cruder and more evident goal than dissecting 'heart failure' at the molecular level.

In a review paper, Bonham et al. (2016) examined the relationships between 'race' and biomedical advances. They claim that "Self-identified race does not predict the genotype or drug response of an individual patient". Over and above the question of whether human races are meaningful biological categories (see Boniolo and Lorusso, 2008; Alan, 2013), the authors put their finger on biologically and clinically relevant aspects regarding the fact that 'prescribing medications on the basis of race oversimplifies the complexities and interplay of ancestry, health, disease, and drug response', and that PM could provide a solution. As an example, Bonahm and colleagues discuss the first 'race'-based pill—BiDil (a combination of two existing drugs - isosorbide dinitrate and hydralazine)—a treatment for heart failure in self-identified African American patients. Unfortunately, this way of grouping individuals neglected many genetic, epigenetic, and social factors influencing both disease development and response to treatment within the same self-identified ethnic group. As another example of this erroneous approach to ethnic diversities, the authors report the first pharmacogenomics lawsuits focusing on differences in genotypes (Wu et al., 2015). In 2014, a Hawaii attorney filed a lawsuit against Bristol–Myers Squibb and Sanofi–Aventis for their failure to disclose that carriers of certain CYP2C19 alleles treated with Plavix (clopidogrel bisulfate, which is prescribed to prevent coronary events and stroke) were either at risk for major complications, or no response at all. The drug had been proposed for the entire Hawaiian population, ignoring the fact that about 56% of the islanders are East Asians, Native Hawaiians, or of other Pacific descent, all frequent carriers of the alleles in question.[11] Put in other words, we no longer need to rely on self-identified human subpopulations but rather on what PM offers, that is, the possibility to consider genetic and epigenetic set-ups and environmental conditions. Again, nevertheless, the rose is not without thorns ....

Hypertrophic cardiomyopathy (HCM) is a cardiac pathology that usually develops the second decade of life. HCM results from a mutation in one of nine genes giving rise to a mutated protein in the sarcomere (Maron et al., 2012). Genetic tests can therefore detect the mutation and diagnose risk of HCM, which is the most common cause of sudden death among the young.

Regrettably, a report published in the *New England Journal of Medicine* warned that the utility of genetic tests may be limited by the lack of diversity of people included in the underlying genetic databases used to assess risk (Manrai et al., 2016). The studies that had established the links between the mutations and HCM did not include data from a substantial number of African Americans in the control groups. The authors concluded that some variants would not have been linked to HCM, had the database contained more genetic information from African Americans. According to the authors of the study: 'The provision of false genetic information to a patient, such as when a patient is incorrectly informed that one of his or her variants is causal when in fact it is benign, can have far-reaching adverse consequences within the family', with the result that both patients and their relatives might drastically change their lifestyles without any good scientific reason.

It should be underlined that these disparities are not related to poor access to healthcare but to the fact that, historically, genetic studies have been conducted on certain populations on the assumption that they were representative of the whole population. The work by Manrai and colleagues was probably the first to show that ethnic-biased genetic data can lead to misdiagnosing certain genetic pathologies, or to proposing new potentially dangerous or ineffective drugs. The problem, it would seem, is not just one of control groups belonging to different ethnic groups, but that pharmacogenomics and other Precision Medicine's strategies are founded on well-understood genotypes. This means that there is a lack of inclusion of genetically diverse populations.

---

[11] Other examples of the fallacies of the self-attributed belonging to a racial or ethnic group are given in Jackson (2008), Ng et al. (2008), Rotimi and Jorde (2010), and Leite (2011).

A review by Ramos et al. (2012) evidences this point, citing the report by the Centers for Disease Control and Prevention[12] warning that even if

> pharmacogenomic information has been added to over 70 drug labels, . . . the studies on which label information are based have mostly focused on European populations. Meanwhile, African populations, who have the greatest genetic variation resulting in more haplotypes, lower levels of linkage disequilibrium, more divergent patterns of linkage disequilibrium and more complex patterns of population substructure, are grossly underrepresented in the genomic studies that inform pharmaceutical guidance. The result is that clinicians may rely too heavily on data obtained from Europeans to make clinical decisions for Africans and other non-European populations.

New projects, however, such as the International Hap Map Project,[13] the 1000 Genomes endeavors,[14] and even the new indications contained in the Precision Medicine Initiative, can surely ameliorate the situation. Nevertheless, today the question remains open and we are in a condition of global injustice.

Conclusions

Precision medicine raises open questions concerning data deluge, biomarker validation, new clinical trials, patients-in-waiting and too much medicine, unknown genotypes and ethnic discrimination, cost of new treatments, global injustice, patients' lack of awareness of the proper meaning of the statistical and probabilistic results, and so on. Nevertheless, this does not imply that we should have a dismissive and totally negative approach to Precision Medicine.

Precision Medicine bears great potential for improving our quality of life and the quality of life. Yet, this requires working from a philosophical point

---

[12] Centers for Disease Control and Prevention, Genomics and Health Impact Blog. Medications for the Masses? Pharmacogenomics is an Important Public Health Issue (2011), https://blogs.cdc.gov/genomics/2011/07/21/medications-for-the-masses-pharmacogenomics-is-an-important-public-health-issue/ (accessed 21 October 2021).

[13] International HapMap Project, https://www.genome.gov/10001688/international-hap map-project (accessed 21 October 2021).

[14] 1000 Genomes Project, https://www.genome.gov/27528684/1000-genomes-project (accessed 21 October 2021).

of view too, both offering ethical analyses and methodological suggestions whenever required, and willing to help in the training of the new generations of clinicians (Boniolo and Campaner, 2020).

Acknowledgements

I would like to express my sincere thanks to Margherita Brusa and Yechiel Michael Barilan for their passionate work and the help they gave to all of us during and after the meeting 'The Revolution of Personalized Medicine. Are We Going to Cure All Diseases and at What Price?' at the Pontifical Academy of Sciences. I would also like to acknowledge the incredible panel of participants with whom I spent a wonderful intellectual time. Thanks also to Raffaella Campaner for her insightful comments and the anonymous reviewers for their suggestions.

5

The precision paradox

How personalized medicine increases uncertainty

Henrik Vogt

According to one definition, precision medicine is 'the tailoring of medical treatment to the individual characteristics of each patient' (National Research Council, 2011). This concept, here treated synonymously with personalized medicine, is as old as medicine, and was for a long time a humanistic medical ideal (Tutton, 2014). However, since the 1990s and the Human Genome Project, a new conceptualization of precision medicine has arisen from within biomedicine. The idea here is to improve diagnostics and treatment, and especially risk prediction and disease prevention, through knowledge of the biological characteristics of each individual case (Feero, 2017).

What 'biological characteristics' means in personalized medicine has been changing. According to an earlier understanding of the concept, such as the one espoused by George Engel in the biopsychosocial model, 'biological' may be understood as a holistic concept[1] (Engel, 1977; Vogt et al., 2014, see also Canby Robinson, 1939; Gibson, 1971). It incorporates everything about the human organism or person—from the molecular to the personal and cultural.

With the new, technologically driven conception appearing in the 1990s, 'biological characteristics' came primarily to mean the genome, and personalization meant treatment according to a person's DNA (Langreth and Waldholz, 1999). During the last twenty years, as science has advanced, and as new biological complexities have been unveiled, the definition of biological characteristics has expanded (Vogt, 2016; Chakravarti, 2011; Shendure, 2019). The US Precision Medicine Initiative defines precision medicine as an

---

[1] Biology does not have to be—and has not always historically been—a reductionist endeavour. That the term 'biological' has come to be associated with cells and molecules is primarily a result of the dominance of molecular biology. The biopsychosocial model rested on the organismic (or holistic) biology of Ludwig von Bertalanffy (see, e.g., Engel, 1977).

Henrik Vogt, *The precision paradox* In: *Can precision medicine be personal; Can personalized medicine be precise?*. Edited by: Yechiel Michael Barilan, Margherita Brusa, and Aaron Ciechanover, Oxford University Press. © Oxford University Press 2022. DOI: 10.1093/oso/9780198863465.003.0005

approach that 'aims to understand how a person's genetics, environment, and lifestyle can help determine the best approach to prevent or treat disease'.[2]

This more comprehensive outlook is enabled by a host of new technologies for gathering and analysing information (Hood and Flores, 2012; Vogt et al., 2016). In addition to genomics, there are other 'omics' technologies for measuring bodily constituents, such as proteomics (protein), transcriptomics (RNA), metabolomics (metabolites), epigenomics (epigenetic markers), and microbiomics (micro-organisms colonizing the body). There are clinical sources of big data, including imaging technologies and electronic health records. An increasing array of biosensors or wearables that monitor physiology and environmental exposures has emerged (Xu et al., 2019). Organoids and organ-on-chip technologies, where patient-derived cells are used to form mini-models of individual physiology, are also emerging (Zhang, 2019). From tailoring medicine to a person's fixed biological markers (the genome), precision medicine has expanded to the continual surveillance of biological markers and digitized data on all other bodily processes (Torkamani, 2017).

In the discourse on precision medicine, the term 'precision' may also hold different meanings. It is used in a broad, colloquial sense to mean both 'accuracy', 'precision', and 'certainty' (National Research Council, 2011; Hunter, 2016). In other words, precision medicine intends to reduce epistemic uncertainty, which according to one broad definition is 'related to our knowledge, of the state of a system' (Djulbegovic, 2011). Ultimately, precision medicine seeks 'the right treatment, at the right time, every time in the right person' (White House, 2015).

The argument in this chapter is that precision medicine represents a paradox with regard to its promises: while seeking to increase precision and certainty in medicine, it might also create more imprecision and uncertainty. Three aspects of this paradox are presented below.

Uncertainty about what works: the return of the art of medicine

A fundamental, ever-present question in medicine is how we know what a symptom, finding or biomarker means and, relatedly, how we know that our interventions work in the particular patient. To know what works, we have

---

[2] https://medlineplus.gov/genetics/understanding/precisionmedicine/initiative/

to compare the individual case, presenting before us now, with a model that provides knowledge about what will happen next. In science, one has preferred generalizable truths validated by rigorously executed experiments (a universalist stance). But clinical medicine addresses health problems of individuals, who present unique cases (Canby Robinson, 1939). Consequently, medical practitioners have stood at an intersection between natural science and what is often called the art of medicine. In this chapter, 'the art of medicine' means the creative act of knowing and doing what is right for the individual, when approached as a whole person (see Marcum, 2008, ch. 1). The art of medicine is the application of scientific knowledge to clinical practice, by integrating such knowledge with other sources of information, notably clinical epidemiology, (patho)physiology in both a narrow and a wide biopsychosocial sense, case histories, the doctor's experience, and patient narratives (Marcum, 2008). It involves a comparison of this integrated knowledge with corresponding information about the individual patient coming from various levels of existence (from the molecular to the social and spiritual). Hence, the art of medicine comprises methods associated with the natural science, the social sciences and the humanities (Marcum, 2008). Until recently, this sort of integration was considered 'personalization' or 'humanization' of medicine.

The tension between the science of generalizable and standardized medicine and the art of personalization has been acknowledged throughout medical history (Tutton, 2014). William Osler (1849–1919), the American founding father of scientific medicine, stressed that, 'If it were not for the great variability among individuals, medicine might as well be a science and not an art' (cited in Tutton, 2012). In the same vein, the American physician George Canby Robinson (1939) stated in his book *Patient as a Person*:

> Recounting experiences in medicine has in a sense gone out of fashion, and the style today is to report measurements, calculations, controlled experiments, and data that can be treated statistically, with the expectation that facts are being acquired which fit into general laws of nature or from which general principles may eventually evolve. These are the ultimate objects of scientific research . . . . In the study of man as an individual it is impossible to apply the tenets of controlled experiments, as each individual presents a unique situation and the nature of the study does not allow strict objectivity. . . . Each person represents an experiment without control.

What Canby Robinson wrote can still be read as a challenge to anyone who wishes to turn personalization into something scientific. Although there have been attempts to incorporate clinical expertise and personal experience into evidence-based medicine (Sackett, 1997), EBM can be seen as a reaction precisely to what is known as the 'art of medicine'. It is the comparison of groups within a population for ascertaining causal effects that distinguished epidemiology from the older medical predictions based on (patho)physiology and evidence from case histories (Broadbent, 2017). One of the leading figures in the development of clinical epidemiology, Sir Austin Bradford Hill, stated:

> Our answers from the clinical trial present . . . a group reaction. They show that one group fared better than another, that given a certain treatment, patients, on the average, get better more frequently or rapidly, or both. We cannot necessarily, perhaps very rarely, pass from that to stating exactly what effect the treatment will have on a particular patient. But there is, surely, no way and no method of deciding that (cited in Djulbegovic, 2011).

Contrasted to Canby Robinson's statement, Bradford Hill presents a grand epistemological challenge to anyone promising that an integrative art or science can truly individualize decisions of care. But in a paradigmatic twist, this is exactly what is being promised in contemporary biomedicine as I will discuss next.

Today's biomedical precision medicine is often presented as being in opposition to evidence-based medicine. According to Beckmann and Lew (2016), for example,

> Precision medicine needs to be contrasted with the powerful and widely used practice of evidence-based medicine, which is informed by meta-analyses or group-centred studies from which mean recommendations are derived.

While precision medicine is also often presented as compatible with evidence-based medicine (Beckmann and Lew, 2016), some of its leading visionaries see it as changing perceptions of what evidence and medical knowledge is:

> We are entering an era in which we can measure those factors that randomized controlled trials have considered dark matter in the information space

of patient-specific traits and have randomized away. We can determine the multi-causality network of each individual patient. The arrival of the 'omics technologies—genomes, epigenomes, transcriptomes, proteomes and metabolomes—is about to shine light into this dark matter of patient-specific causality. This offers an opportunity to unite the 'unscientific', but patient-focused alternative medicine with the 'scientific', evidenced-based best practices, or one-size-fits-all medicine (Kauffman et al., 2014).

What these authors call 'alternative medicine' roughly corresponds to the personalization and the 'art' of clinical practice that antedated evidence-based medicine. Their idea seems to be that the computations introduced by biomedically based personalized medicine will replace—and render scientific—this patient-centred practice. This vision of turning what has been art into science can be seen an underlying promise of precision achieved by personalized medicine.

There is a daunting epistemological problem with this promise of uniting the art of personalization with the generalizability of science. As biomedicine seeks to personalize, the number of people who are deemed to have a health problem moves towards 1 ($n = 1$). It then becomes hard or impossible to determine what works for patients by using statistically-based methods that compare populations, for example randomized controlled experiments. This is especially true when the number of variables that are measured in each individual increases (Klauschen et al., 2014), and the inclusion of more variables is precisely what is needed to map the 'multi-causality network of each individual' (Kauffman et al., 2014).

As precision medicine is associated with scientific biomedicine, its protagonists will likely not want it associated with art. They will first attempt to generate, find, and employ as much relevant, statistically-based evidence as it can in determining what is best in the individual case. The main strategy in uniting science and evidence-based medicine with personalization will be stratification, which means aggregating cases that are similar and then comparing them statistically to controls (Beckmann and Lew, 2016). New statistically-based, empirical methods like umbrella and basket trials are also proposed to provide new footing for evidence-based medicine (Schork, 2015). However, as biomedicine traverses into the scientifically unchartered terrain of the individual, the number of subjects that are defined as having a problem may become so small that statistics cannot be used in any method regardless of the effect size. This is the first aspect of what I propose as the precision paradox. As one cannot use the statistically-based methods that have been central to evidence-based medicine, it becomes uncertain what findings mean, how we determine what works in medicine and how we are to agree what is documented and what is not. This creates uncertainty about what will be the tenets of evidence-based medicine in the future.

One strategy that is proposed to mitigate this situation is the n-of-1 trial where the patient is his/her own control (Schork, 2015). In n-of-1 trials the patient and doctor can be double-blinded and there are washout periods between administration of placebo and active medicine in the same patient. Such studies do present one promising way forward for precision medicine in aligning with the tenets of evidence-based medicine, but for ethical reasons (periods where the patient gets a placebo pill may mean death in, e.g., cancer patients) and practical reasons (they demand substantial resources and it is difficult to follow periodic diseases) they may in many instances not be feasible.

It is important to note that the uncertainty that will be generated by precision medicine is not new. Considerable uncertainty in medicine comes from applying class probability (predictions of what works derived from populations) to case probability (predictions about the individual patient) (Djulbegovic, 2011). Such uncertainty can be seen as having previously been 'hidden away' in population-based guidelines. In precision medicine, this uncertainty is 'released', in the sense that it becomes clearer how difficult it is to know what works in the individual and how to actually cope with this apart from issuing guidelines that provide general rules of practice.

One of the first instantiations of such uncertainty in knowing what findings mean and what treatments work in individuals is currently faced by tumour boards in hospitals providing cancer medicine around the world. These boards, which vary in their composition and routines, consist of experts who convene to discuss individual cases (Hoefflin et al., 2018). Even when clinical studies are not available, tumor boards perform judgments by creatively matching mechanistic [patho]physiological knowledge to genetic information about the patient's tumour, other kinds of molecular information, imaging and clinical information (Ree et al., 2017). Given what Tonelli and Shirts describe as the 'impossibility of conducting standard population-based analyses when the number of individuals with a specific variant is very small', precision medicine thus means a return to mechanistic (pathophysiological) reasoning and case-based reasoning (or case histories) as a basis for a 'broader concept of medical knowledge' (Tonelli and Shirts, 2017). Because there exists to date no rigorous scientific criteria for incorporating case histories and mechanistic reasoning in processes of reliable decision-making, this integrative reasoning was considered part of the art of medicine. Such criteria are now being sought (Tonelli and Shirts, 2017), but we cannot know if they are even possible to develop in a way that enables us to reliably assess whether personalized treatments work.

While personalized medicine may mean a return to reliance on physiological models and case-based reasoning, it will differ from the past by its reliance on algorithmic processing of large amounts of data. However, it may in one sense also remain quite traditional because the ultimate face of this new medicine may be narratives, stories about individuals whose lives have been changed by personalized medicine. Leroy Hood, a pioneer personalized medicine, states in (Gibbs, 2014) that . . .

> We hope to develop a whole series of stories about how actionable opportunities have changed the wellness of individuals or have made them aware of how they can avoid disease.

Although the 'stories' Hood envisions are filled not with words and experiences of patients, but of billions of data points, we see an appeal to use case histories as evidence.

In sum, in precision medicine we see the art of medicine rearing its head just as a tide of techno-scientific biomedicine promises to bring science to the individual.

Bog data and big noise: complexity and precision

The second aspect of the precision paradox is that the greater accuracy of complex models may be inversely correlated with greater precision. In their personalized medicine manifesto referred to above, Kauffman et al. (2014) argue for building "very high-dimensional 'state spaces' of thousands of pieces of environmental, genetic, epigenetic and physiological data—so-called big data". Furthermore, Kauffman et al. argue:

> Personalized medicine strives to exploit precisely the information that is randomized away by RCTs[3] by studying one thousand patients or more. Instead of the false certainty obtained by neutralizing patient-specific

---

[3] Randomized Controlled Trials.

variation in studies that include as many people as possible, personalized medicine seeks, ideally, a full molecular profiling of an individual patient. . . . The prerequisites for this approach are high-dimensional measurements that leave no or little unaccounted-for factors that would have to be randomized, and mechanistic knowledge that allows predictive models based on the high-dimensional data obtained for each individual.

Kauffman and colleagues propose that big data measurements that are extremely precise in the sense of being granular ('full molecular profiling'), as well as high dimensional can describe the causal complexity.

As pointed out above, in the biomedical literature the term "precision" bears three different, even if somewhat overlapping meanings: accurate, precise and highly certain (National Research Council, 2011; Hunter, 2016). 'Accuracy' usually means that the models and diagnostic tests validly reflect the reality of the health problem, with a low degree of bias. 'Precision' exists when a model gives the same result every time with a low degree of variance or noise (Fortmann-Roe, 2012). However, precision may also hold a technical meaning distinguished from repeatability or reliability, namely 'refinement in a measurement' (Lexico, 2020).

Precision, both in the sense of greater accuracy and increasing refinement in measurement, may come into conflict with precision in the sense of reliability—and certainty.

Many years before the advent of precision medicine and big data science, computer scientist and originator of fuzzy mathmathics, Lotfi Zadeh, argued that precision, in the technical sense of mathematically precise input for computers, is seldom what medical decision-making needs. Rather, decisions often must be based on imprecise—or 'fuzzy'—information. According to Zadeh (1972),

In general, complexity and precision bear an inverse relation to one another in the sense that, as the complexity of the problem increases, the possibility of analyzing it in precise terms diminishes'.

Zadeh called this principle the 'the principle of incompatibility'. Systems biologist Olaf Wolkenhauer and his colleagues have called it 'Zadeh's uncertainty principle' and "the biggest problem that any approach to mathematical modelling in biology faces" (Wolkenhauer et al. 2004). According to these authors:

Overly ambitious attempts to build predictive models of cells or subcellular processes are likely to experience the fate of historians and weather forecasters—prediction is difficult, especially if it concerns the future, and these difficulties are independent of the time, amount of data available or technological resources (e.g. computing power) thrown at the problem.

The tension between complexity and precision is also captured by the 'bias variance dilemma', also known as 'the curse of dimensionality' (see Figure 5.1) (Bellman, 1957; Fortmann-Roe, 2012; Alyass, 2015; Barbour, 2019). In general, high predictive value of medical tests (low total error) requires an underlying model that incorporates the complexity of the biological system with high accuracy (validity) and with precision (reliability or repeatability). Up to a certain point, the inclusion of more variables (dimensions) and measurements leads to greater accuracy (reduced bias) and thus a better predictive value. However, with additional variables and measurements, the variance then starts to increase quickly, resulting in reduced precision (or increasing noise). Beyond a certain threshold, the level of noise exceeds the reliability of the signal (Fortmann-Roe, 2012; Alyass et al., 2015).

The second aspect of the precision paradox is that, although medicine is now capable of describing the body with highly accurate and precise measurements, precision is not necessarily compatible with describing the biological complexity of individuals. Big data is sometimes thought of as

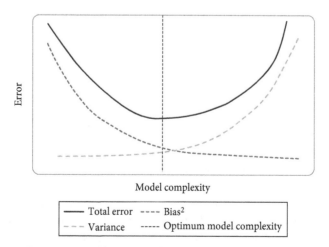

Figure 5.1 Bias-variance dilemma: an illustration of the tension between model complexity and precision (Fortmann-Roe, 2012; used with permission).

enabling ever more precision and certainty. But, to coin a new term, big data may sometimes become 'bog data'—data that drags the scientist into noise and uncertainty like a quagmire. The more you struggle with its complexity, the more it pulls you under. In fact, big data may paradoxically mean the advent of 'big noise'.

Delivering medicine all the time, at every time, in every person

The third aspect of the precision paradox concerns the preventive part of the precision medicine vision.

In Ancient Greek philosophy, *kairos* referred to the opportune moment. It was used by Hippocrates to describe the art of treatment given at the right time and to account for individuality and specificity (Eskin 2002). It seems that the promise of contemporary biomedical precision medicine is also a form of *kairos* medicine: To 'deliver the right treatments, at the right time, every time to the right person' (White House, 2015). Like the health counterpart to precision warfare, which also often seeks to strike enemies preemptively before they cause 'symptoms', precision medicine promises to target only the right risk factors, while letting 'innocent bystander' conditions go.

This is the paradox: In order to be there at the right time, in the right person therapeutically, precision medicine needs to be there all the time, in everybody in terms of risk profiling and diagnostics. To take the war metaphor further: Its precision warfare against disease and risk can only be established through a state of medical surveillance, which monitors each and every one constantly—a medicalization of human life (Vogt et al., 2016). Because no surveillance is wholly precise, and there are limits to risk prediction, precision medicine thus risks being there at the wrong time in the wrong person.

Preventive precision medicine aims to pick up risk conditions and early disease at an early time-point. An EU funded consortium, for example, highlights 'the possibility of identifying the origin of certain diseases at the earliest possible time during the life-course to enable early secondary prevention' (PerMed, 2015). While intuitively appealing, this drive towards diagnosis at 'the earliest possible time' is a leading driver of overdiagnosis (Pathirana et al., 2017). Unlike false-positive results, overdiagnosis is defined as the detection of conditions that are actually seen as abnormal or suspicious by medicine, but which are not destined to become significant health problems as in causing symptoms or death (Brodersen et al., 2018). This occurs when diagnosed abnormalities will either disappear spontaneously without medical attention, or progress so slowly that the person dies from other causes before the abnormalities cause symptoms. Overdiagnosis is a problem because harm may result from diagnostic labelling and worry in itself, and because it leads to overtreatment with harmful side-effects (Brodersen et al., 2018).

Even though precision medicine promises to alleviate overdiagnosis (Hunt and Jha, 2015), early detection followed by early intervention can be beneficial only when abnormalities progress to symptomatic disease. However, the natural histories of biological abnormalities show great variation, and so does the competing risk of dying from other causes than those screened for. As the ability of different technologies to observe the human body at different levels increases, it becomes, in a sense, increasingly 'transparent'. A full 'transparency' of the body through time, including genome sequencing, will also mean that we will be able to 'see' everything that medicine at any time defines as abnormal, suspicious or of unknown significance (Vogt et al., 2019).

The challenge for precision medicine then becomes to predict which of the abnormalities will lead to symptoms or death. Those registered abnormalities that will not lead to symptoms or death, will represent overdiagnosis (Vogt et al., 2019). Because complex systems are hard to predict, or may show fundamental unpredictability (Firth, 1991; Smith, 2011), precision medicine's capacity to do this is limited (Alyass et al., 2015; Wolkenhauer et al., 2004, see also Nebert and Zhang, 2012).

Thus, precision medicine faces the risk of creating a deluge of findings of low risk and risks of unknown significance. One might argue that massive surveillance is the path towards *kairos* in terms of treatment given at the right time; nonetheless such findings will require follow-up and often prompt treatment with predictable side-effects. If we invoke the metaphor medicine as war against diseases, the quest for precision warfare against disease, risks turning into a form of medical carpet-bombing. The enterprise of limiting side-effects by precision comes at the price of all-inclusive surveillance. An ever-present medicine is the opposite of *kairos*.

Conclusion with some ethical implications

In this chapter, I have highlighted three aspects of what has been dubbed the precision paradox. Precision medicine, despite its promise, may also increase uncertainty in medicine and become *imprecision medicine*. I want to conclude by noting that these three aspects are clearly interrelated:

The first aspect of the paradox is that, just as medicine promises to bring personalization into the fold of natural science, it becomes harder in some instances to know what findings mean and if interventions work. This concerns the cases where *n* approaches 1. This first aspect is related to the second aspect of the paradox, as it becomes harder to know what will work in the individual when the number of variables that are measured in each individual increase. The second aspect concerns the increasing imprecision and noise—and thus uncertainty—associated with increasing complexity of data and models. As evident in Barbour's (2019) definition of 'the curse of dimensionality', this problem is related to the first aspect of not knowing what works for the individual:

> Our intuition regarding '*average*' is rooted in one-dimensional thinking, such as the distribution of height across a population. This intuition breaks down in higher dimensions when multiple measurements are combined: fewer individuals are close to average for many measurements simultaneously than for any single measurement alone.

The third aspect of the precision paradox is where precision medicine risks ending up medicalizing everyone, everywhere, all the time, every time. Medical risk assessment will become an everpresent part of conscious life, a state that the philosopher Hans-Georg Gadamer defined as incongruent with health (Gadamer, 1996).

The medical monitoring state will likely label everyone with health problems and lead to a tsunami of overdiagnosis. This will be worsened by the first and second aspect: When it becomes hard to predict what findings or risk factors mean because *n* approaches one and the complexity of models increases, it becomes harder to know which of the many abnormal findings will actually lead to symptomatic disease.

In sum, genuine personalization must help the person focus on important issues and not worry about the rest. But precision medicine risks becoming imprecise, clouding the medical gaze with lots of distracting noise and distracting people from their non-medical life goals.

Acknowledgements

I want to acknowledge that the term 'precision paradox' was coined by professor Jan Helge Solbakk, Center for Medical Ethics, University of Oslo. I also want to acknowledge that parts of the description of overdiagnosis above draws on previous writings in collaboration with my co-authors in Vogt et al., 2019: John Brodersen, Claus Ekström, and Sara Green. Finally, I wish to thank my PhD supervisor professor Linn Getz as some of this chapter builds on my PhD thesis (Vogt, 2017).

6

Genomics and precision medicine

Through a different prism

Farhat Moazam

"Each of us touches one place
and understands the whole in that way
. . . if we went in together
we could see it"
       Elephant in the Dark, Rumi (Persian Sufi poet, d. 1273 CE)

Introduction

The prism I will use to view precision medicine and genomics-centred healthcare paradigms is of a physician with longstanding interest in bio-ethics, who, following education in the USA, has practised medicine for many years in private and public (government) health sectors of Pakistan. In the ongoing discussions and flurry of publications on these topics there is a paucity of voices from low- and middle-income countries, whose populations and health professionals will inevitably be affected by the genetic 'revolution' as it seeps through their borders. While there has been exponential increase in information about the human genome, much of this territory remains uncharted with novel discoveries around each corner and new questions emerging about how best to utilize this knowledge for the benefit of humanity. Mevlana Jalal ud Din's couplet, that to truly 'see' the elephant in a dark room requires a multifaceted endeavour, seems an apt metaphor for where we stand today in relation to genomic medicine and its promises. My observations will be familiar to other low- to middle-income country physicians but may also resonate with health-related professionals in high-income countries involved in the care of citizens who face socio-economic and other hardships not too dissimilar to those residing in countries like Pakistan.

Farhat Moazam, *Genomics and precision medicine* In: *Can precision medicine be personal; Can personalized medicine be precise?*. Edited by: Yechiel Michael Barilan, Margherita Brusa, and Aaron Ciechanover, Oxford University Press. © Oxford University Press 2022. DOI: 10.1093/oso/9780198863465.003.0006

The Human Genome Project, spearheaded by the National Human Genome Research Institute (NHGRI) in USA and completed in 2003, has served as the catalyst for research leading to useful insights into genetic contributions to human illnesses and providing tools for improving management of some diseases (Human Genome Project, 2020). However, the view that genomics medicine should form the basis for radical transformation of research and clinical practice in the future remains contested. Supporters of P4 (predictive, preventive, personalized, participatory) systems medicine, resting in genomics, term it a healthcare 'revolution' that will integrate 'an individual's genetic, molecular, cellular, organ and social networks into an overall 'network of networks'', and which in their opinion is 'with us already' (Hood and Auffray, 2013). Others remain sceptical of such claims, questioning whether precision medicine's promises of improving risk prediction and health outcomes, lowering treatment costs, and contributing to enhanced public health are realistic (Joyner, 2015).

Two-thirds of the 7.7 billion human inhabitants of the planet live in either low- to middle-incomes or low-income countries (Trading Economics, 2020). Pakistan, a South Asian low- to middle-income country, is the sixth most populous nation in the world with a population of 208 million (Worldometers, 2020). It has a per-capita income of US$1,640 as compared to US$40,000 in the UK and US$65,000 in the USA (Wikipedia, 2020). Borders of low- to middle-income countries are increasingly porous to biotechnological advances in medical practice and multinational collaborative research, for which they often serve as hosts. The movement across borders, subsumed within the broad mantle of 'scientific progress', is largely in a North to South, West to East direction. The impetus comes from various quarters including healthcare professionals who train overseas and return home to practice their specialties, progressive privatization of medical care in the face of ineffective national health services, and governments unable/unwilling to finance research within local academic institutions and interested in encouraging foreign investments (Moazam, 2006; Bhatt, 2015; Mondal and Abrol, 2015).

These factors are impacting both the nature and focus of clinical practice and the goals and trajectory of research in developing countries. Many of the advances connected to modern medicine are available only in the private, pay-for-service health sectors, accentuating existing ethical challenges to fairness, equity, and justice in access to healthcare as well as to professional and national responsibilities towards patients. In the case of research, developing countries provide large, illiterate, treatment-naïve populations, willing English-speaking researchers, and less stringent oversight mechanisms, all of which can substantially reduce costs and time for data collection for drug and clinical trials funded by overseas agencies including the pharmaceutical industry (Glickman et al., 2009).

The landscapes in developing countries can be quite different from those that exist in countries that are currently in the forefront of scientific advances. A proposed shift to precision medicine and genomics as the dominant paradigm for future medical practice and research requires deeper and more inclusive attention to social issues and ethical trade-offs that may become necessary at local and global levels. An important component of the 1990 Human Genome Project was the incorporation of the Ethical, Legal and Social Implications (ELSI) Research Program which meant to identify and address ethical, legal, and social implications of genomic research and the potential for misuse of resultant technologies (Ghr.nlm.nih.gov). This was an acknowledgement that humans are more than biological amalgamations of genes and molecules in some universal sense, and that illness, health, and wellbeing are experiential and contextual, shaped by complex cultural, social, and environmental influences. Ethical and social concerns highlighted in ELSI thirty years ago remain relevant today at national and global levels. A recent editorial in the *American Journal of Medical Genetics* called for attention to the implications of genomics in a global context 'with a special emphasis on low and middle-income countries' (Lázaro-Muñoz and Lenk, 2019). Although the editorial focus was on psychiatric genomics, these concerns are equally relevant to other areas in medicine.

In this essay I will begin with brief observations about what I see as the individualistic ethos and language that underpins genomic medicine and which contrasts with the ways in which many societies in the world construct the individual and her/his relationship to others. Using my experiences as a clinician, I will follow this with what I believe will be the implications for developing countries should genomics medicine come to define the standard of care in clinical medicine and the direction of future research.

On ethos and language

'I am the master of my fate, I am the captain of my soul'
Invictus, William E. Henley (British poet, d. 1903)

'The link that binds the individual to the society a mercy is . . . He is a jewel threaded on [the community's] cord'
Mysteries of Selflessness, Allama Iqbal (Asian philosopher and poet, d. 1938)

The study of regional poetry can be an effective means for insight into values that peoples consider important and the many different ways in which societies shape the self and its relationships to others, including family and society. Henley's couplet quoted above presents the ideal of an autonomous individual in control of her/his destiny influenced by Enlightenment philosophical traditions in which the rational individual is the primary moral and decision-making unit of society and external authority is suspect. The language of the autonomous individual is a recurrent theme in publications and consumer advertisements pertaining to genomics medicine, the 'empowered' individual and the public who take 'personal control' of their healthcare needs (Juengst et al., 2012b).

Genomics is presented as the science that will 'use a person's clinical, genetic, genomic, and environmental information to tailor a treatment plan that will maximize efficacy and safety for that individual' (Wolpe, 2009). A 'more individualized, molecular approach' is offered for a future in which extensive biological information will be accessible to individuals through electronic records' (Collins, 2015). Genomics science is the wave of 'individualized' healthcare that will replace contemporary medicine in which 'large cohorts of individuals [are treated as] uniform' by physicians who are in control (Institute for Systems Biology, 2014). The emerging systems biology programme promises to transform the 'passive patient' into an 'engaged consumer who takes ownership of his or her own health' (https://www.faceb ook.com/p4Mi.USA/info). Similarly, the language of direct-to-consumer campaigns in a rapidly growing consumer genomics market in America presents having information about one's genome as personal empowerment, thus conflating information with knowledge of a complex, still-evolving field (Juengst et al., 2012b).

The notion of an individual in control of her/his life contrasts with the worldview of societies that are predominantly collectivistic in nature, family centred, and that live profoundly interdependent lives shaped by duties and obligations rather than by individual rights. Allama Iqbal, revered and among the most quoted poets of the Asian subcontinent, reflects this in his couplet above, in which the individual and community are seen as blessed when woven together, a view that your life cannot be yours alone, unconditionally. Within these social norms prevalent in many Asian and African countries, major decisions including those pertaining to health and illness are arrived at collectively by the family rather than by individuals. Among less privileged socio-economic classes the interdependencies are augmented by necessity as earnings are pooled for support of large numbers of family members. My research in Pakistan to understand how medical decisions are taken, using organ donation and transplantation as paradigms, has repeatedly revealed this to be the case with few exceptions; in effect you are your family and your family is you (Moazam, 2006; Moazam et al., 2014).

It is also interesting to note the language adopted by the United Nations General Assembly and its 193 member countries in 2015 when it agreed on seventeen goals, the Sustainable Development Goals (SDG), to be achieved by 2030. As health is considered a public good, the primary moral principle was stated as 'leaving no one behind' with the responsibility to ensure 'healthy lives and promote well-being for all at all ages'. The language is one of global solidarity and responsibilities with a focus on addressing mundane ailments such as maternal and infant mortality rates, '[providing] access to medicines for all', and supporting research and vaccines for diseases that 'primarily affect developing countries' (www.un.org/development/desa/disabilities/envision2030-goal3.html). In my opinion, this is ambitious but necessary in a world with shrinking resources and growing inequities in healthcare access, a situation from which now even HICs are no longer immune.

On research implications

'We were glad they took the blood . . .
they come because treatment is possible, that is obvious.
Why would they come if there is no treatment?'
    Pakistani genetic research subject (Sheikh and Hoeyer, 2018)

Converting the promises of precision and genomics-based medicine into reality will require funding for research to develop large biobanks of genetic profiles, biological specimens and health-related data from patients and healthy volunteers to establish 'big data'. As expressed by supporters, there is a need to develop 'personal data clouds of billions of individuals' that can be 'mined' (Ghr.nlm.nih.gov).

Credible, empirical public health studies over several decades reveal that the greatest impact in improving population health and reducing health disparities, whether in HICs or LICs, is the result of high-quality primary care and prevention programmes rather than high-end research and technological advances (Sankar and Cho, 2004; West et al., 2017; Chowkwanyun et al., 2018).

It is likely that with shrinking national and institutional resources for research, funds may have to be prioritized for genomics-related research over the 'traditional' forms of studies. Perhaps a harbinger to this is evident in Ronald Bayer's and Sandro Galea's report that in 2014 the funding for research proposals in America that included 'the words gene, genome, genetic was fifty percent greater than for those with the word prevention' (Bayer and Galea, 2015).

The US National Institutes of Health (NIH) in America has initiated an 'All of Us Research Program' with funding running in the millions. The aim is 'to gather data over many years from one million or more people living in the United States' in order 'to develop individualized care' (National Institutes of Health, 2020). The Hastings Center in New York recently announced its collaboration with NIH to help with the enrolment of 75% of the one-million cohort from 'underrepresented groups' in the country. This will be done via engagement with a health centre serving 'primarily Latino and African American patients', the laudable goal being to try to understand why socially disadvantaged groups have been 'hesitant to enrol in biomedical research' (Hastings Center, 2019).

While discussing ethical and social challenges of personalized genomic medicine, Juengst et al. (2012) note that the goal and methods will inevitably depend on the interpretation of 'key stakeholders, especially those who dominate the movement'. High-income countries have a relatively better informed public awareness of research risks and potential for discrimination based on genetic profiles, and stringent national regulations to be satisfied before initiating research, clinical trials, and introduction of new drugs into the market. Low- to middle-income countries, however, do not pose these hurdles to research in light of large, poor populations with high illiteracy rates and lax or missing ethical oversight and regulatory mechanisms at institutional and national levels (Petryna, 2007; Jafarey et al., 2012). It is inevitable that enthusiastic low- to middle-income country researchers and governments interested in attracting foreign funding will be ready for collaboration with genomic research proposals to assist in developing genetic biobanks and repositories with patient tissue samples and information.

An informed, comprehended and voluntarily given consent by participants is universally accepted as the lynchpin of ethical human subject research. In discussions about the Precision Medicine Initiative being undertaken in America, it was noted that ability of research participants to provide an informed consent may be compromised due to limited knowledge about genomics 'especially among underserved communities and those with lower educational attainments' (Sabatello and Appelbaum, 2017). The challenges to obtaining ethical informed consents in developing countries will be even greater when participants are not only poor and illiterate but also when they do not understand the difference between research and therapeutic interventions, and implicitly trust physicians (researchers in white coats are presumed to be doctors) as those who know what is in their best interest (Moazam et al., 2009). Language barriers and use of local translators to explain the nature and purpose of genetic research will compound the difficulties.

Beginning in the 1990s, there has been a steady increase in multinational, collaborative human subject research, including clinical trials funded by industry, in many low- to middle-income countries. There is growing interest in genomics among researchers in developing countries, and multinational, collaborative, internationally funded genetic research is on the rise. A survey by one of my graduate students revealed fifty-two publications (in PubMed and open access journals alone) from January 2017 to December 2018, involving single-gene or genome-wide research conducted in rural Pakistani sites.[1] Of these, 88% were international collaborations, over half were funded entirely through international agencies and/or consortia, and due to lack of local facilities all blood and tissue samples were stored outside Pakistan. Her review also revealed that twenty publications included a blanket statement that 'an informed consent was obtained' without details of the process and nature of information conveyed to research subjects. It is also noteworthy that while international, collaborative research in Pakistan requires prior ethical review and approval by the National Bioethics Committee (NBC), none of the fifty-two publications indicated that this was done. I am a member of the NBC and it is not uncommon for the committee to learn about international, collaborative research projects which, inadvertently or otherwise, were not submitted for ethical clearance. Countries such as Pakistan that lack robust national oversight systems and enforceable regulatory mechanisms make them attractive sites for multinational research conducted on human subjects. It is also noteworthy that although international, collaborative research requires prior approval by the National Bioethics Committee of Pakistan, none of the fifty-two articles mentioned that this had been done (Ethical Clearance, 2020).

[1] The student, a molecular biologist, holds a MA in Bioethics (Class of 2019) from the Center of Biomedical Ethics and Culture, SIUT in Karachi. Her survey reveals that most studies focused on the genetic basis of cardiovascular diseases and inherited genetic disorders in children.

In a study conducted in rural Pakistan by Danish researchers to review the informed consent process used in previous collaborative genetic studies, the primary motivation of families in consenting to blood samples being taken from their children was a belief that this would lead to a cure for their disease. 'I didn't have anything to decide . . . maybe they'll find a cure . . . so we said that's OK'. Others consented because somebody they knew from the village 'convinced' them that the samples would be used 'for cure purposes' (Kongsholm et al., 2018). Another sociological study using interviews with Pakistani genetic research subjects to explore the concept of 'trust' revealed similar aspects of therapeutic misconception on the part of families. Nonmedical researchers were perceived as foreign doctors who would help cure their children based on blood samples collected from them. 'They said they were going to do some tests in Germany . . . Then they would try to take our kids over there for treatment' (Sheikh and Hoeyer, 2018).

Precision medicine will require production of drugs targeting specific genotypes and subsets of patients by a market-driven pharmaceutical industry which relies on making profit. Pharmacogenetics for subsets of patients will be a phenomenally expensive venture for an industry that relies on recouping its investments through production of generic blockbuster drugs. Developing countries with large populations are already used for stage 3 and 4 clinical drug trials by the industry, and extending these to genetically targeted drug trials would not be surprising. Whether benefits that accrue from genomics-related research undertaken on the populations of low- to middle-income countries will be affordable to those people, or available in their countries, is open to question. What is more likely to occur is the transformation of the most disadvantaged populations into biological entities, mere means to an end for developing large data banks. In other words it would be an example of the Golden Rule that appeared in a 1967 *Wizard of Id* comic strip—'whoever has the gold makes the rules.'

On clinical practice implications

'Only professionals who stand on the firing line or in the trenches
can really appreciate the moral problems of medicine'
James F. Childress, *Practical Reasoning in Bioethics* (1997)

One of my most enduring memories after returning to Pakistan following training and practice in America for several years was operating on a one-week-old neonate born with a complex trachea-oesophageal fistula.[2] I conducted the surgery in a state-of-the-art operating room in a hospital in Karachi equipped with the latest technology no different from what I was using in the USA. The infant's discharge from the hospital a few days later, feeding normally and gaining weight, was a moment of triumph for my surgical team. The little boy died eight months later from infectious diarrhoea because the village in which his family lived did not have access to clean drinking water. Such indelible experiences are not uncommon for physicians working in low- to middle-income countries, where most of the planet's population lives. For the more reflective, such cases can raise troubling questions about the much-heralded 'progress' of biomedical science and the focus of what constitutes healthcare.

Among the reasons given for supporting personalized and systems-based medicine approaches is that these will serve as a corrective for what is described as a 'disease-oriented' and therefore 'dysfunctional' current (American) health system. Those involved in 'old fashioned' medicine are being curtly advised to 'Lead, follow, or get out of the way' (Snyderman and Yoediono, 2008). The USA's overall expenditure on healthcare is the highest in the world if measured relative to gross national product. Yet, it has poorer health indicators than many less affluent countries, and ranks 24th in achieving health goals set by the United Nations (Sainato, 2020). Many argue that the genomics approach is too narrow in its understanding of a complex problem, is insufficiently attentive to its ethical and social effects, and at best will provide no more than 'marginal improvements at high cost' (Campos-Outcalt, 2008). Henrik Vogt and his colleagues, among them Norwegian general practitioners, argue that the P4 systems medicine paradigm in particular will also change the basic ethos of medicine from humanistic and holistic to one that is 'technoscientific' in nature. In their opinion, 'buzzwords such as "genomics," "big data," "digital health" and "personalized" or "precision medicine"' form a nexus within which human life will be defined in 'biomedical, technocratic terms as quantifiable and controllable' (Vogt et al., 2016). My experience as a clinician has taught me that it is short-sighted to believe that the human experience is quantifiable.

---

[2] A congenital condition in which the child is born with a connection between the trachea (the tube going down to the lungs) and the oesophagus (the tube through which food passes into the stomach). During feeding the milk enters the lungs of the baby instead of going down to the stomach, leading to pneumonia and death.

For physicians, patient-centred medical care which tailors the most effective therapy for patients through attention to individual variations and relevant social factors is not a new concept. Genetic testing, biomarker screening, and eventually disease-targeted drugs can serve as additional tools towards achieving this goal. However, accepting these as the new international standard for clinical practice will accentuate challenges to provision of equitable, non-discriminatory access to healthcare in the face of existing socio-economic disparities. Genetic testing for certain diseases is now available in developing countries but primarily within private, fee-for-service health domains affordable to few. In Pakistan, genetic screening for breast cancer means out-of-pocket expense for tests ranging from US$160 to well over US$200. The cost of one vial of Herceptin for oestrogen receptor-positive breast cancer is around US$900 with injections required every three months for at least one year. According to a colleague in one of the largest cancer hospitals in Pakistan, a one-year survey of their breast cancer patients revealed that only '1% [of suitable patients] could afford Herceptin due to the prohibitive costs' (personal communication).

The 2013 report by the European Commission-led Genetic Testing in Emerging Economies (GenTEE), surveying the status of genetic screening in emerging economies, revealed the absence of universal health cover in many countries, placing the tests beyond the reach of most patients who cannot afford such out of pocket services (Christianson et al., 2013). If genetic testing becomes the yardstick for 'standard of care' in the future, another layer of injustice will be added to existing healthcare inequities between the 'haves' and the 'have nots', the affluent few who can afford private medical care and the disadvantaged many who must rely on public services. The chasm in discriminatory healthcare practices will widen even further by the fact that many developing countries allocate low percentages of their GNP to public healthcare services and are increasingly ceding a national responsibility to the burgeoning, for-profit, private health sectors which are inaccessible to the majority of populations.

Effective use of the specialized nature of human genomic and pharmacogenomic information requires sufficient genetic literacy among healthcare professionals in the front lines of patient care. The subject of genomics is currently not part of the curricula in medical colleges and postgraduate training programmes in most countries including some within the western hemisphere. In the absence of this knowledge, 'customizing' care that will work best for a patient is open to misinterpretation and miscalculation of genetic results, increasing potential risks to patients (Moazam, 2006).

The situation regarding professional genetic illiteracy is direr in developing countries. The GenTEE report notes that most physicians are unable to recognize the genetic basis of diseases, as well as the need for genetic referral and counselling of patients. Compounding this problem is the fact that the number of certified genetic counsellors ranges from 0.06 to 0.2 per million population as compared to the recommended 3 to 5 counsellors per million population (Christianson et al., 2013). As far as I have been able to determine, Pakistan with a population of over 200 million people currently has, at the most, three trained, certified genetic counsellors.

The case of genetic illiteracy appears little different when it comes to members of the general public. Sabatello and Juengst (2019) note that in America, even while 'genomic information is ever more infused in public discourse' and through direct-to-consumer marketing, understanding of genomics and its role in health and illness remain low. In low- to middle-income countries genetic illiteracy and misinformed beliefs and myths about hereditary traits are even more common among the public. In my own practice, parents have sometimes informed me that children get their 'blood' only from the father's side of the family and yet, paradoxically, on other occasions have inquired whether a particular congenital condition in the child was passed down through the mother's blood.

Genetic information, unlike other personal data, does not belong to the patient alone as it may hold implications for other family members. The ethical importance of privacy and confidentiality related to genetic information, in research and in clinical practice, and the difficulties in how these can, and when they should or should not be maintained, are currently being debated in different forums. The issues of privacy and confidentiality related to genetic information will be compounded in low- to middle-income countries characterized by hierarchical, patriarchal, and patrilineal family structures in which three or more generations live under one roof, and healthcare decisions are taken collectively. Unlike routine investigations that provide data specific to the patient's illness, genetic tests can reveal sensitive information with implications for the larger family. This will compound the complexities of ethical medical decisions which will require balancing the welfare of the patient versus responsibilities of the physician to her family. In the absence of trained genetic counsellors and when genetic information is inadequately and/or incorrectly explained by treating physicians and is poorly understood by patients and their families, severe repercussions are likely for the most vulnerable members of the family. Female members especially could be at high risk for blame and stigmatization if genetic findings are revealed to the family.

Concluding thoughts

'The merchant serves the purse,
The eater serves his meat; . . .
Things are in the saddle,
And ride mankind'
   Ode, Ralph W. Emerson, American philosopher-poet (d. 1882)

Better understanding of the human genome may undoubtedly help to provide additional tools for better management of patients. But genomics is one tool among many within healthcare; it is not the silver bullet, as it is sometimes portrayed, which can transform healthcare including by '[reducing] health disparities' between resource-rich countries and low- and middle-income countries (Rehman, Awais, et al., 2016). In 1993, Eric J. Cassell, a physician, wrote about the 'irresistible spread of technology', which, despite its key role in producing medical economic inflation, lures physicians, nations, and indeed the public, with 'the immediate', 'the unambiguous', and 'the pursuit of certainty' (Cassell, 1993). As a professor in a leading university in Pakistan when I was invited to give a talk in medical institutions, I was sometimes taken for a round in the hospital. Even in public hospitals, frequently short-staffed and lacking basic drugs for patients, I was struck by the focus on showing me the latest biotechnological equipment and machines which the institution had acquired.

The lures Cassell identified over two decades ago live on in some of the hyped promises of precision medicine and genomics-based 'big data' medical systems, albeit in forms that are more scientifically sophisticated, more expansive, and significantly more expensive. Currently there is an absence of greater inclusive discussions at national and global levels—not limited merely to scientists who are producing these tools, but including those who will be wielding these tools and those upon whom they will be wielded, as well as an assessment of the trade-offs that may be required. Without such discussions, we run the risk—to paraphrase Ralph Waldo Emerson—of science in the saddle riding mankind.

There is compelling scientific evidence that disparities in health access and care of populations, nationally and globally, are related to social determinants rather than the genetic make-up of individuals. Physicians and public health personnel working at the forefront of healthcare understand that health and illness are experiential and subjective, culturally and environmentally modulated, and that humans exist within immensely diverse 'local worlds' that defy biological universalistic notions (Kleinman, 1997). Yet the vision for a genomics-based medical system for the future of medicine is an attempt to atomize the complex human condition into quantifiable entities that can be measured mathematically through algorithms, loaded into 'clouds of data', and managed through 'a stream of new drug targets for the pharmaceutical industry' (Flores et al., 2013).

Half a century ago, Ivan Illich wrote about the progressive 'medicalization of life' by an expanding and increasingly commercialized health industry and a 'pharmaceutical invasion'. He argued that medicine is displacing human values and human understanding of health and illness by assuming 'the authority to label one man's complaint a legitimate illness', and to 'declare a second man sick though he himself does not complain'. Today we appear to be standing on the threshold of medicalizing the entirety of human life by extending medicine's reach from patients who are suffering from disease to those who are healthy, i.e. patients in waiting, as commercial ventures. In their vision of future healthcare, Hood and associates of the Institute of Systems Biology describe the design for a scientific wellness system based on 'dense and dynamical data clouds' and 'quantitative metrics' which will 'improve the wellness' of healthy individuals. The authors estimate that once operational this 'new industry' will eventually 'exceed the "disease" industry in market cap' (Hood et al., 2014).

Precision medicine and its variations raise important ethical challenges at professional, national, and global levels about fairness, equity, and justice in healthcare, the balancing of the rights of individuals versus duties and responsibilities to society. It raises profound questions about what should be the direction and focus of scientific research and medical practice in a world with limited and unevenly distributed resources, how and who will decide the 'all' in the 'all of us', and the kind of world which we wish to leave to future generations. These matters require collective wisdom; they are too important to be left to molecular biologists and genetic scientists alone. I will end by returning to Mevlana Rumi (1997), the Muslim thinker and Sufi with whose elephant metaphor I began. Rumi states in another poem that 'having the idea is not living the reality of anything.'

7

Personalization, individuation, and the ethos of precision medicine

Yechiel Michael Barilan

Introduction

Medicine has developed as a response to diseases that are threatening and disruptive, often justifying extraordinary measures and a sense of emergency. Some feats of personalized medicine fit this paradigm, helping, for example, to find the most effective and least toxic chemotherapy for patients with cancer. However, in its more ambitious versions, personalized medicine is not limited to coping with health crises, but it is a way of life and a shift in the way people perceive human nature and personal goals. This chapter discusses the transition from responsive medicine, which is focused on specific conditions and threats, to medicine as a 'personalized way of life' aimed at the optimization of health or 'wellness' (Vogt et al., 2016; Juengst and McGowan, 2018).

Every scientific practice needs an 'attendant myth' to give it stability, direction, and identity (Shapin, 1996, p. 95). Beyond the moral implications of specific words and metaphors looms large the power of medicine to impose its specialized language on people's experiences and values (Frank, 1995, p. 6). This chapter explores a web of meanings, associations, and metaphors that sustain the discourse on 'personalized medicine'.

I will argue that the ideas and promises of personalized medicine, its 'attendant myth' and 'specialized language' embody a secular and high-tech version of ancient and medieval natural law morality, and that the technologies of personalized medicine are rooted historically in morally charged practices. The allure of personalized medicine masks risks to the human person as an agent in society and as an autonomous patient of medicine.

Yechiel Michael Barilan, *Personalization, individuation, and the ethos of precision medicine* In: *Can precision medicine be personal; Can personalized medicine be precise?*. Edited by: Yechiel Michael Barilan, Margherita Brusa, and Aaron Ciechanover, Oxford University Press. © Oxford University Press 2022.
DOI: 10.1093/oso/9780198863465.003.0007

The grand narrative of personalized medicine: secularized and pragmatic rendering of natural law morality

The era of artificial intelligence (AI) and information technology (IT) has established itself upon the computation of large amounts of data coming from a diversity of sources (Russell, 2019; Lepore, 2020; Larson 2021). The key idea behind personalized medicine is that biotechnology and 'big data' informatics transform personal data into practical information about one's health. Further, personalized medicine, and the role it has granted genetics, are rooted in the idea that the genome is a set of *instructions*. A leading metaphor is that the genome is 'the book of life' (Karow, 2001), a privileged text that encodes the rules for the good life. Medicine 'reads' the embodied codes of genomics, proteinomics, and other markers, and issues instructions for a healthy life.[1]

The scientific jargon speaks of genes as prescriptions for proteins, and the genome as a set of prescriptions for protein synthesis and regulation. The moral and medical idiom is about a new kind of healthcare that seeks to decipher *personal instructions*. This transition from instructions as a metaphor in scientific explanation to instruction as a moral imperative is borne out by James Watson's address to the American Academy of Arts and Sciences in 1990, referring to the recently launched Human Genome Project: 'What's more important than this piece of instructions?' (Fortun, 2015, p. 41). In those days, scientists referred to non-coding DNA as 'junk'. When it was later discovered that 'non-coding' sequences encode different kinds of 'instructions', the 'junk' metaphor was abandoned.

Personalized medicine is founded on the assertion that the genome stands for the essence of the person, at least as a biological creature. Not only does this assertion build on the uniqueness of every human individual, but it also dovetails every individual in a grand matrix of family and ethnic relationships. This geneticization of personal identity overlaps with a universal and ancient conception of persons as knots in webs of kinship. Whereas in traditional human societies kin status and cultural norms predicate one's social horizons, personalized medicine aims at foretelling one's biological horizons—his or her risk profiles in terms of disease and response to treatment, urging people to structure their lives in consideration of their anticipated biological future.

---

[1] The special status of the genome as 'The Book' might account for the widespread abhorrence at 'editing' the genome.

Because every person has a unique genome, and because science teaches that genetic codes are the primary regulators of the organism, the implicit message is that every person has a unique set of instructions regarding his or her health. Moreover, even though all humans share over 99% of their genome, the attention to personal variations shifts the weight of identity and personal choice from the common to the unique, from 'we' to 'me' (see Dickenson, 2013), from health as a fruit of interactions with the human and non-human environment to health as compliance with inner instructions, inscribed on the hearts of every cell of the body.

This narrative portrays the genome as an Aristotelian 'primary mover' in the sense that personalized medicine beholds the genome as the cause (efficient cause) and purpose (telos) of human life, at least its biological prosperity (see Rommen, 1947, pp. 49–50). Aristotle's notion of 'natural justice', which 'everywhere has the same force and does not exist by people's thinking' (Nicomachean Ethics, 11348b18) does not distinguish between fact and norm. Personalized medicine reiterates this approach by a smooth transition from scientific theories about pathogenesis to good healthcare. The metaphor of the 'selfish gene' and the predominance of genetics in evolutionary theory underscore further the message that the genome is the person's fate, destiny, and laws of action.[2] Personalized medicine does not portray genetics as a sealed fate, but as a code that enables scientists to infer instructions for predicting and intercepting.

Consequently, the genome has become a new and secular version of natural law.[3] Natural law provides the 'epic of knowledge' that supports the

---

[2] The evolving concept of the 'hologenome' encompassing all DNA and RNA encoding in a person's 'system' calls into question the relevance of a life-long, fixed master-book of one's life (Rosenberg and Zilber-Rosenberg, 2018). Chimeras and mosaics fit the instruction metaphor and even with the uniqueness of the individual, but less with the notion that genetic uniformity overlaps embodiment (the uniformity of body-person—see Barilan, 2012, pp. 220–3).

[3] By this term I refer to a school of thought, attributed mainly to Augustine's and Aquinas's elaboration of Aristotle and the Stoics (see Adams, 1998, pp. 133–8). According to doctrines of natural law, reflection by human rationality on human nature, as part of the natural order as a whole, elicits anthropological and normative insights about human goals as a basis for people's good and virtuous lives. According to natural law, nature must refer to an essence that has a goal or accomplishment. In personalized medicine, a healthy existence is the optimization of human nature. Modern biology has marginalized teleological reasoning. personalized medicine discourse seems to resurrect natural law thinking in an ironic exemplification of Heinrich Rommen's original title of his masterly book on natural law—'Natural Law Eternal Return' (Die Ewige Wiederkehr des Naturrechts, Leipzig, 1936) (Rommen, 1947). The return of telos is especially problematic when it is carried by statistical correlations.

narrative and legitimizes the social face of personalized medicine by appropriating the ultimate goal of medicine in the era of calculations (see Lyotard, 1984, p. 27).

Whereas in natural law theory, common rationality guides normativity, in genetics, rationality helps us decipher nature's instructions. Whereas naturalistic ethics relies on reflective thinking, personalized medicine relies on sophisticated calculations. Being risk-oriented, personalized medicine acts as a fortune-teller, fulfilling the archaic human wish to live in awareness of the future. Within this framework, knowing one's horizons of risks amounts to self-knowledge that is key to the good life. It also shifts human attention from facts and choices to risks and opportunities as quantifiable realities.

Whereas in natural law theory, rational reflection on the nature, goods, and goals (*telos*) of human life instructs us how to act, genetics is the set of instructions indicating how to self-govern as risk managers. The shift from information about one's biological markers to prescription of behaviour and the good life is the natural fallacy of personalized medicine.

In the coming sections, I will first explore the historical tensions between individuation and personalization. I will then show that personalized medicine's individuation by means of computational matrices of relative risk might undermine personal agency in relation to one's health, and further undermine people's capacity to act as agents in the democratic governance of healthcare.

Individuation, identification, and personalization

Individuation is a method for pinpointing a human individual apart from others. Pure individuation is about capturing an individual, without knowing something substantial about him or her. Fingerprinting is a method of pure individuation because the fingerprints do not correlate with health, ethnicity, creditworthiness, or any other cultural context. Fingerprinting is quite efficient as an individuator because it is highly specific, it does not change over time and place, and it is easily obtainable, stored, and processed.

Identification is a method for connecting a human individual with a method of individuation. In modern societies, facial photos are common methods of identification. Facial recognition is the default method of identification because it is the least expertise-dependent and because facial interaction is also a universal and ancient means of personalization. Human brains are especially attuned to the recognition of faces and facial differences.

Personalization is always stable and global in the sense that it refers to the whole human being (globality) throughout his or her life (stability). It is about referring to and interacting with a human individual, distinct from any other, and valued as a human person. Consequently, personalization in relation to the person as a biological body cannot be disconnected from the person as an agent.

Because individuation is context-specific, there could be more than one method of individuating a person (e.g. a passport number and a social security number individuate the same person). There could be no relationship whatsoever between individuation by means of biometrics and individuation by means of date and place of birth. By contrast, absence of interconnections between one aspect of personalization and another is depersonalizing. It is alienation. Consequently, this chapter is exploring the risks of alienation within a healthcare system whose moto is personalization.

All human societies use abstract names (culturally created symbols) to perform the triple task of individuation, identification, and personalization. A personal name serves as a means of individuation (my name is different from others', especially if it is combined with a surname/name of a parent), of identification (I know my name and respond to it; people who know me directly link my appearance, habitus, and voice with my name) and of personalization (it is usually meaningful and given by the parents, rather than an arbitrary number).

People tend to detest individuation and identification by bodily features. Branding, identification by fingerprints and identification numbers are considered as potentially dehumanizing. When doctors talk about 'the diabetic near the door' they individuate and identify in a dehumanizing manner. Representing context-specific individuation and identification as 'personalization' could be offensive and highly depersonalizing.[4]

In complex societies, individuation and identification depend on diverse technologies, which are more precise and more amenable to large-scale processing than personal names and facial photos. The scale of large societies and of globalization harbour a tension between personalized means of

---

[4] People tend to adapt fast. One early response to the telephone was that 'the dignity of talking consists of having a listener, and it seems absurd to be addressing a piece of iron' (*Popular Science*, March 1878, p. 626). In the era of social distancing, much human communication is electronically mediated, with an as yet unclear impact on our sense of personalization and respect for persons.

individuation and identification on one hand, and their precision and accuracy on the other hand. In large scale societies, digitization is necessary for precision in individuation.

Because every human being is born with a unique 'genome', because this genome is part of his or her body, because the genome is perceived as a book of instructions, and because technologies of genomic individuation and identification have become rife, many people conclude that the 'genome' might also be a means of personalization.

Some versions of personalized medicine inspire this conclusion. Moreover, whereas individuation and identification are discrete (either a person matches the ID number/photo or not), genetic data is also contextual. It can inform about family relations, ethnic background, and about relative risk (e.g. risk of developing major depression, diabetes, etc.). Consequently, the construction of personal identity upon genomics and other 'omics' is a novel kind of individuation.[5] It individuates people on scales of risk relative to other people. It ties individuation and personalization to comparative matrices. For example, a pioneer of population genetics, Robert Plomin, individuates a person by showing his combination of risks. Plomin locates this person on the 22nd percentile in relation to bipolar disorder, 33rd percentile to major depression, 39th percentile to Alzheimer disease, 85th percentile to schizophrenia, and 94th to 'educational attainment' (Plomin, 2018, Fig. 11). Arguably, there is no other person with this combination of risk profiles. (If there is one, it is possible to add scales of risk so as to exclude any chance of overlap.)

Genomics and 'personalized medicine' depend on information technology (IT) and 'big data' science. Algorithms applied to combinations of genetic markers (e.g. alleles), proteinomics, and other markers create the 'personalization' in personalized medicine. This horizontal interconnectedness (many markers culled at a single time-point) often comes with longitudinal surveillance (the collection of ongoing personal data) so as to create an image of unity and comprehensiveness, thus rendering the impression that 'personalized medicine' captures the person as a unified human agent, from conception until death. Eric Topol, an enthusiast of personalized medicine, beholds such comprehensiveness as 'deep medicine', whose bottom rock is genetics (Topol, 2019, p. 16).

---

[5] 'Identical' and 'conjoined' twins challenge the geneticization of personal identity (Barilan 2003).

[Deep medicine] deeply defines each individual (digitizing the medical essence of a human being) using all relevant data. This might include all of one's medical, social, behavioral, and family histories, as well as one's biology: anatomy, physiology, and environment. Our biology has multiple layers—our DNA genome, our RNA . . . .

People tend to believe that digitization and its big-data processing can 'mine our soul, excavate deep into the biological, to peel away the psychological' (Harcourt, 2015, p. 127).

Personalized medicine's reach into the genome yields the impression of capturing the essence of the person, rather than haphazard, obscure, and mutable properties. Personalized medicine reads the code of instructions thus reflecting to the person his or her digital *doppelganger*, his or her 'data double' (Haggerty and Ericson, 2000). Personalized medicine can act both as a confessor and as an instructor. Non-compliance with the confessor's instructions seems like obstinate refusal to listen to one's own truth, a truth rooted in a contemporary version of biologized natural law. Thus, personalized medicine appears irresistible. Reflection on the evolution of systems of individuation might expose some conceptual and ethical issues challenging this allure.

A historical background to 'personalized medicine'

As the biblical story on the Judgment of Solomon illustrates, problems of individuation and identification challenge the most brilliant minds (I Kings 3:16-28). However, the story of personalized medicine begins with a different kind of individuation. It is not merely about individuation *from* others, but also individuation *relative to* others. Additionally, it is individuation in the form of comparative risk stratification. It might be a positive risk (i.e. chance) such as capacity to perform a certain task, and it might be a negative risk, such as sudden death.

In the second half of the nineteenth century, the emergent business of life insurance heralded the science of risk stratification among groups. Actuarial calculations enabled insurers to rank risk according to medical and demographic information. 'Classing' people by risk-groups developed into 'targeting' people by their buying or voting powers and tendencies (Bouk, 2015). These methods enabled 'personalization' of a particular 'treatment', such as the calculation of premiums, marketing campaigns, or disseminating propaganda. They have developed in other directions too. Actuarial algorithms assessed the risk of recidivism and thus influenced decisions on parole (Harcourt, 2007). The leading value was efficiency—efficient pricing of insurances, efficient production at work, efficient crime control. In the case of personalized medicine, the efficiency is defined by health-indices and cost-effectiveness of medical interventions.

Until World War I, capacity to self-support by one's own labour was taken as evidence of fitness for military service. In 1916, the UK army established the National Service Board, which conducted medical exams of over 2.5 million candidates to service, classifying them into four categories of health. The trove of findings that came from the mass examination of all drafted men brought forth the notion of preventive medicine in the form of medical examinations of apparently healthy people. Such large-scale endeavours involved automated methods, and, ultimately, the use of computers (Reiser, 1978, pp. 216–26). Screening medical exams that were offered by life-insurance companies in 1911 as a 'life extension service' turned into life-saving medical care for all. Prevention by means of early detection became 'the first duty of medicine' (Gladstone, 1932).

The pretext of national emergency in war facilitated the creation of databases of personal medical information. Disaster relief, a synonym of national emergency, was the legal justification for the creation of the Social Security Number (SSN) system (Landis Dauber, 2012, pp. 122–3). Half a century later, promoters of the expansion of newborn screening similarly invoked the value of 'saving lives' in emergency situations, attributing it to the 'rule of rescue', as justification to mass screening of all babies (Cookson et al., 2008; Brusa and Dickenson, chapter 8). Today, the accumulation of medical data on everybody by healthcare providers has become the norm.

Registries of birth and deaths and the allocation of personal identification numbers created a bureaucratic shadow that would accompany people throughout the life cycle. Items such as father's name could be sensitive private data. Children born out of wedlock are stigmatized by the absence of a father's name in the registry (Robinson, 2015, ch. 5). Religious and ethnic affiliation may lead to discrimination, claims for empowerment, and assertions of equality (Igo, 2018, pp. 64–88).

The advent of 'scientific management' (Taylorism) of labour during the early twentieth century introduced the idea that productivity and safety at work can and should be 'personalized' by means of minute observation and calculation. The lure of scientific management was embodied in three promises. It offered work that was safer and more fulfilling; it offered employers higher productivity; and it offered the public (workers included) cheaper products (Kraines, 1960). In a similar vein, 'personalized medicine' promises better and safer healthcare to patients and lower costs of healthcare (by spending only on those who are likely to respond). Whereas 'scientific management' approached persons as agents of bodily movement and motion (Karns Alexander, 2008, pp. 11–12), the rise of white-collar labour shifted attention to cognition and sociability. Consequently, personality testing and psychological evaluations became rife during the second half of the twentieth century. They extended the growth of Taylorism in bureaucratic formats. Employers sought to 'personalize' recruitment; many workers wanted personalized advice regarding career choices (Igo, 2018, ch. 3; Erme, 2018).

Computers ushered in the 'science' of 'big data', which allows fast processing of registries and other data banks. This meant the capacity to produce knowledge out of 'big data'. It became possible to 'feed' the data to computers and come up with novel insights, even cast predictions. The expectations of computers were huge. In the late 1950s, pioneers of Artificial Intelligence assumed that 'if they could collect enough data about enough people and feed it into a machine, everything, one day, might be predictable'. Since then, computers appear as 'the new seers, electronic prophets, diviners of data' (Lepore, 2020, p. 15).

The 'big data science' of advertising, 'predictive policing', and political campaigning can also predict medical prognosis and response to treatment. However, predictions at the group level (populations) pertain to events (e.g. mortality in the coming year); predictions at the individual level pertain to risk (e.g. the risk of sudden death in the coming year).

It is noteworthy that some 'big data' operations have so far failed to reach fruition, for example, attempts to capitalize on predicting trends in the stock markets. There is no way of telling which kind of 'predictive–preventive' operations are solvable by algorithms that process 'big data'.[6] Algorithmically solvable problems must rely on numbers and equations that stand for health and wellness, thus invoking the Goodhart Law: When we use a good numerical marker for optimizing a system, the marker is not useful anymore (Crystal and Mizen, 2003). For example, if we take vitamin D levels as a marker of health and plan healthcare accordingly, then, all efforts will be directed at vitamin D, ignoring side effects. The only way out of this pitfall is a continuous re-evaluation of the markers, the optimization scheme, and the

---

[6] The P=NP problem (Wigderson, 2019).

ultimate goal (e.g. wellness). This cannot be done unless we understand the algorithms.

The 'war against terrorism' was another episode of 'national emergency' justifying, in the eyes of the state, invasion into privacy by means of 'big data' predictive science (Harcourt, 2015). Personalized medicine has never appealed to state coercion and invasion. Rather, it has appealed to the free choice of people, who care to improve their health and to help promote the health of others. People are called in to participate in such enterprises, such as the UK 100,000 genome project. The notion of 'participation' is the willingness to share personal data and to accept surveillance. Personalized medicine evokes individualism and solidarity alike.

Whereas personalized medicine's narrative speaks the language of natural law, its historical evolution appeals to pragmatic utilitarianism and humanity's weakness for fortune telling. Notwithstanding the multiple promises of personalized medicine—to alignment with human nature, to individuality, to efficiency and to solidarity—some serious moral difficulties arise stemming directly from the construction of 'personalization'. In the next sections I explore some of these difficulties.

Individuation and depersonalization

There are endless ways for individuating a human being. For example, it might be the case that the only human who was born in Ajaccio and who died in St Helena is Napoleon Bonaparte. It might also be possible to individuate Napoleon by his denture, DNA, and fingerprints. However, addressing an autonomous human being as a person is, first and foremost, addressing his or her self-perception, set of values and responsibilities, which may range over many domains, such as career, family, hobbies, friends, religion, community, and politics. The prioritization of these values and responsibilities is a dynamic personal complex, reflecting the course of a person's life. The autonomous human person embraces responsibilities and creates priorities in any given circumstance. It follows that 'personalization' that addresses the human as a person aims at the embodied human individual and his or her experienced life and locus of responsibilities.[7] This is what physicians meant

---

[7] This chapter relies on the occidental conceptualization of the person as a unified conscious self, which forms a locus of responsibilities and subjective rights (i.e. a source of claims in the name of the person, such as his or her desires). There are alternative theories of the self, such as the 'partible' person (Strathern, 1988). In the case of incompetent humans (babies, significant

when they spoke about personalized medicine before the era of IT (Robinson 1939; Gibson, 1971; reviewed by Shorter, 2015).[8]

When marketing, employment, medicine, and national security talk about 'personalization', these social domains address one or a few selected aspects of the person. Marketing personalizes by desire and buying power; scientific management personalizes by mechanical efficiency in teams; the criminal system and national security agencies care about public safety, not about the person in focus.

Alas, approaching a person from one direction *depersonalizes* other personal aspects. Personalized marketing risks diminishing the person to a consumer[9]; personalizing labour risks diminishing the person to a machine. Personalized medicine risks diminishing the person to a patient and, perhaps even only to the set of risks to health and life. Personalization by psychiatric risk may diminish personalization by alternative panels of medical risks.

Diminishment of the person by consumerism, productivity, or health implies lack of integration of other aspects of human flourishing, bypassing the agent as a hub of diverse responsibilities. This might generate alienation between the human person as a dignified autonomous agent in society and the system of personalization used, including personalized medicine.

Even though medicine personalizes through the prism of health and disability, many portray and perceive personalized medicine as seizing the essence of the person and his or her calling. We have become habituated to surveillance and personal monitoring, to their use in labelling people, to instructions handed down by the apparati of the 'therapeutic state', and to social claims based on the labels and on compliance with instructions (e.g. Medicare, social services, food stamps, probation) (Polsky, 1990). Large-scale operations of assistance have justified their incursions into liberty as measures of exceptional emergency; the more underprivileged the person, the deeper and more comprehensive state policing have become (Bendich,

---

neuropsychiatric impairments), personalization addresses the person as a locus of socio-cultural relationships of care, and the empowerment of the agent that the incompetent persons might become when he or she grows up or recovers.

[8] As a development of the "patient as the person" school (see Shorter, 2015), the biopsychosocial model is akin to P4 personalized medicine by theoretical reliance on system biology (Bolton, 2014; Tretter, 2019).

[9] There is no evidence that 'personalized' marketing captures the full market agency of a person; even if such an agency exists and captured, it does not capture the whole person as an agent. The same goes for medical and other domains of 'personalization'.

1966). The vision of personalized medicine appropriates practices we have been constrained to accept as promises we should be willing to embrace.

Beyond the risk of *external* depersonalization by means of selective individuation and monitoring, personalized medicine runs the risk of *internal* depersonalization (disintegration and alienation) of people from their capacity to manage their healthcare autonomously. When cultural values and roles (more specifically the 'patient role') clash with personal capacities, experiences, and values, a malignant process of alienation sets in. George Simmel referred to this as 'the tragedy of culture' (Simmel, 1911/1968). In the next sections I will show how the cultural assimilation of personalized medicine's ethos might generate such a tragedy. It might hamper human capacity to deliberate their own healthcare and to execute the actions they might wish to take in its promotion.

The human scale of personalization: understanding and action

The concepts of disease and ill-health are human universals. All known human societies have disease entities (Hocket, 1973). Diseases are composed of explanatory schema, narratives that link descriptions (e.g. a particular skin rash and headache), circumstances (e.g. appear in the winter), mechanisms of causality (e.g. 'because of humidity'), and temporal courses (e.g. resolves within weeks) (Clements, 1932; Thagard 1999, pp. 5–36). Diseases are not necessarily entities, but explanatory models (Good, 1994, p. 53). These explanatory models are anchored in 'somatization', which is a basic idiom for the expression, interpretation, and monitoring of the course of disease and healing (Fábrega, 1997, p. 197). Whereas many disease processes occur beyond human perception, the models of disease must contain *experiential events in the body of the sick person, and there only*—morphological changes, sensations, altered capacities—an overall sense of ego-dystonia, being unhappy with something within the self.

The models must also contain explanatory narratives that integrate these experiences. An astrological chart, a biochemical process, and many other elements of explanatory models do not explain disease and healing unless they interact with somatizations. A fundamental presumption of biomedicine is that it is possible and desirable to have an objective knowledge of the body apart from experiencing the body (Good, 1994, p. 117).

With the rise of modern medicine and large-scale clinical research, people have become accustomed to diseases that nobody feels, such as hypercholesterolaemia and hypertension. Such diseases have become ubiquitous; diagnosing and treating them have become a major part of healthcare. Consequently, people 'suffer' from conditions such as hypertension, diabetes, and osteoporosis, even when the condition is symptom-free. The significant and direct impact of these symptomless conditions on morbidity and mortality justifies diagnosis and treatment as if they were diseases.[10]

Whereas the medical discourse on hypertension and diabetes is anchored in scientifically collectible facts, the discourse on health and wellbeing is formed from past and present cultural life (Pellegrino and Thomasma, 1981, p. 63). It must contain the medical discourse, but it is irreducible to it. Even while referring to them, medical discourse must contain meaningful messages beyond markers labelled as "normal" / "pathological". Optimization of specific health indices is often a nebulous range rather than a definite point of equilibrium (e.g. thyroid function is considered 'normal' when thyroid-stimulating hormone levels range from 0.4 to 4 mIU/L). Even though some cultures harbour ideals of 'super-health' (Alter, 1999), there is no evidence that health is an optimizable good and that its components (e.g. physical and mental) are fully commensurable with each other.

Moreover, optimization of specific processes does not guarantee the ultimate goal these processes serve. Use of AI in "targeted killing" of the enemy may save resources and reduce collateral harm. It does not necessarily lead to winning the war; use of AI to predict recidivism may reduce crime but not promote overall justice (Crawford 2021, ch. 10). In the same vein, successful optimization of healthcare processes does not guarantee better health.[11]

Even if personalized medicine limits itself to an optimized set of evidence-based targets (e.g. TSH levels), the choice of targets might be value-laden. It might be possible that AI may arrive at the desired endpoints (beneficial predictions and interventions) by means of more than one set of markers. One such set may contain six genetic items, three proteinomics markers, and five demographic markers; another set may contain five symptoms, three social markers, and four imaging-based markers. For convenience

---

[10] Since they are not 'diseases' in the ordinary sense of the word, a philosopher of medicine called them 'morbi' (Taylor, 1979, chs 7 and 8).

[11] A designer of an educational website stated, "I think it is legitimately possible for us to create an algorithm that builds personalized catalogue [of on-line courses] that leads you to become the best person you can be" (Friend, 2021).

reasons, personalized medicine will opt for the first set. Consequently, even though there is more than one way to 'personalize' medicine, the version that will eventually mature might reflect and enhance societies' biases, values, and coincidences. It might generate a strong bias towards genetics and towards easy-to-obtain and quantifiable markers at the expense of dynamic, non-digitized and costly-to-obtain data. This will create a bias towards 'disease' (i.e. biological)-related data at the expense of 'sickness' (i.e. social dimensions) and 'illness' (i.e. the felt experience)-related data. Sadly, the last type of data—when patients open their heart and explain their illness and values, is the type most commonly and philosophically associated with 'personalization'.

Moreover, personalized medicine may come up with algorithms that do not factor in information such as 'wheelchair-boundedness' but use proxy data that bypass it, such as proteinomic markers of low physical activity, and markers related to the condition that renders the person wheelchair-bound. Issues such as occupational health hazards might similarly go undetected, even if mathematically incorporated. Considering the immense impact of the social determinants of health on morbidity and mortality, a system that prescribes individual interventions, but no social reforms is likely to miss the human potential for better health, let alone the moral value of social change. Personalized medicine may inspire acceptance of one's genetic horizons; it must not similarly induce acceptance of one's social conditions. This injunction is especially fateful because our genes will not change our social conditions, but social conditions may thoroughly affect gene expression and the ultimate impact of genetics on one's health and quality of life.

Humans tend to set goals for themselves and then to cope with obstacles, including ill-health. The predictive aspirations of personalized medicine threaten to reverse this order and to make people know their 'risk profile' first and make personal choices accordingly.[12] From the perspective of the humanist tradition of the person as an autonomous agent, such a reversal cannot be considered 'personalization'.

Moreover, the notion of 'efficiency' derives its meaning from technology and economics. Detached from these contexts, efficiency in terms of personal

---

[12] Stories of successful prediction and manipulation feed the public discourse on 'big data business', especially those that predict human behaviour, such as social media. Inadvertently, they also inflate the image of this industry, tempting advertisers and investors to spend more money on operations of the kind that have allegedly decided US presidential elections. Notwithstanding justified concerns, we must keep in mind that no proof at the level of rigor required by 'evidence-based medicine' of such capacities of influence has been demonstrated.

goals and personal health is far from clear. There is a risk that the standards of 'health' and 'efficiency' by which the AI of personalized medicine is programmed might influence people's perceptions and values about health and wellbeing. The predictive powers of personalized medicine might also become self-fulfilling (see Lanier, 2013, pp. 166–8). Because genetically based predictions or risk profiles may be cast in infancy, family and society would choose for children. Thus, the constraint on personal choice and an open future is more evident.

The human scale of personalization: capacity to take control over one's own life

Personalized medicine relies on the assumption that knowledge of one's risk and health-related needs is key to health and personal empowerment. Unfortunately, already within the ambience of ordinary medicine, patients' adherence to care is a notoriously intractable challenge. Seventy-five percent of Americans have trouble taking their medications as prescribed. This results in an estimated 125,000 deaths, and causes an estimated US$100 billion in medical bills in the USA each year (Benjamin, 2012). Even in clinical trials, known for the motivation of participants and the extreme commitment of clinicians, adherence rates are in the range of 60% (Osterberg and Blaschke, 2005; Shader, 2018). Awareness of medical risk and proper action guarantees neither understanding, nor empowerment.

Health promotion interventions aim at simple habits that are beneficial to most people, such as high-fibre diet and sport. Yet, even a supportive environment and nudges result in a minor impact on such habits (Arno and Thomas, 2016). In this light, adherence to personalized interventions is expected to be much lower. Moreover, when the number of prescribed interventions piles up, the lower adherence becomes. Up to 50% of older adults do not adhere to at least one long-term medication (Marcum and Gallad, 2012). A leading conclusion is that medicine needs to be 'less disruptive' of people's lives by lowering 'treatment burden' (May et al., 2009).

Polypharmacy, frequent alterations of medications, dietary and activity restrictions, socio-cultural barriers as well as the physiological and emotional burden of disease render chronic and multiple comorbidities an exceedingly difficult task for patients to cope with. Whereas patients may become completely passive during a single intervention such as surgery, cognitive and emotional involvement is necessary for the implementation of interventions that become integrated in people's daily lives. Patients' understanding is essential for the conversion of medical advice into habit.

Normalization Process Theory is a model for understanding, organization, and operationalization of health-related tasks and of transforming them into routine elements of daily life of chronically ill patients. It allows understanding the 'work' and burden involved in being 'a patient of chronic diseases', and how to render this work and burden personally manageable (May and Finch, 2009; Gallacher et al., 2011). Based on qualitative research, this model shows the centrality of personal agency in adherence to care. Long-term adherence requires 'cognitive participation'.

It follows, that to become genuinely 'personalized', personalized medicine should pinpoint a few interventions that are the most important (large effect size in relation to a personal goal) and acceptable to the patient (cost and side-effects commensurate with benefit). Even such interventions suffer from adherence issues. For example, adherence rate to the basic intervention of low phenylalanine diet by patients with phenylketonuria begins with 88% among babies, declining gradually to 65% among adults (MacDonald et al., 2010).

A 'personalized' solution based on predictive algorithms and nudges might improve adherence rates. For example, a leading factor behind non-adherence is fear of medication (Pound et al., 2005). Culling genetic information, and perhaps even bloodborne markers (e.g. adrenaline), AI might detect timid patients and might nudge them to adherence (see Bera et al., 2017). Alas, effective as they might be, the prediction of human behaviour and the capacity to tweak it do not amount to genuine personalization, which is the interaction with the human person as an autonomous subject. The engineering of human behaviour by tailoring interventions at the level of the individual is sophisticated manipulation. It is 'instrumentation', not 'personalization' (Zuboff, 2019, p. 352).

The empowerment of the responsible self, the centre that integrates the subjective (feeling, emotions, streams of consciousness) with the objective (evidence-supported explanatory narratives) is a leading social challenge. This self—'healthy' or 'sick'—should also be an active agent in shaping the social contexts, especially in the regulation of healthcare and the containment of conflicts of interests. Scientists might vouchsafe the validity of personalized medicine in terms of morbidity, mortality, and costs; but personal agency is necessary for aligning 'personalization' with personal values. Herein lies the risk of limiting 'participation' to submission to surveillance, sharing of data, and adherence to prescription, while setting aside effective political participation (See Tilley 2007, ch. 1).

Human societies have self-organized around nation states. Governments set barriers on global participation in diverse activities related to production and consumption such as minimum wage laws and tariffs. Countries create artificial barriers, such as migration laws that bar skilled workers to find employment in foreign countries (Milanovic, 2019, 131ff.; Rodrik, 2011). Participation in the governance of democratic society is mostly limited to citizens. The circle of citizenship usually overlaps the right to employment, to vote, and to benefit from the services of the state, including healthcare. However, if every person's biological profile is unique but essential for medical knowledge, medical science needs 'participation' in the form of global sharing of genetic and medical data. The pool of contributors to healthcare does not overlap the pool of benefactors. In this case, there must be novel forms of regulation of such participation, not allowing a gap between people as sources of medically needed data and people as agents that control the regulation of the benefit generated from that data. Such gaps might alienate people from the form of medicine that calls itself 'personalized'. Additionally, an expectation to share personal data needs be accompanied by standards of data sharing between public and private actors.[13]

Adherence to preventive measures and treatment depend much on circumstances of care. For example, without alternatives, people accept work in unhealthy conditions and forgo drugs they find too expensive. They might also consent to share medical data. Adherence to care, generation of the information needed for medical advice, and the regulation of both are interdependent.

Visionaries of personalized medicine have underscored the centrality of 'participation', without fully understanding its implications, especially its centrality in the creation of ethical operations of care.

Conclusion

The word 'personalization' carries the allure of fundamental values such as identity, self-knowledge, and nature's rules for the good life. The participation

---

[13] Contemporary genomic and similar databases are produced by numerous scientists, technicians, and IT personnel. Public sharing is necessary for epistemological accuracy and quality control of the knowledge produced (Ankeny and Leonelli, 2013).

expected by personalized medicine invokes the values of solidarity, health, and person-centred care. The perception of the genome as a set of instructions infuses additional naturalistic and normative energies into genetic-based personalized medicine. 'Big data science' appeals to people's yearning for knowledge of the future, control of nature, and pragmatic efficiency.

We have seen that reflection on the history behind 'personalization' in personalized medicine reveals that it is closer to individuation in circumstances of emergency than to approaching people as autonomous agents in society. The jargon of personalized medicine and its reliance on IT call into question people's capacity to connect medical diagnosis and prescriptions for action with their experiential lives as political agents, and autonomous patients.

When personalized medicine offers the management of discrete and serious medical problems, its great potential is easier to grasp and these moral challenges are of lesser concern. The more personalized medicine permeates daily life, the more pressing these moral concerns become. The greater is the number of people undergoing continuous personalized surveillance, the more medicalized society is becoming. Rather, this medicalization is technical and computerized. The osmosis of high-tech medicine into every aspect of life becomes the norm, not the exception in need for extraordinary justification, such as saving life. Personalized medicine might direct people's life-choices towards the perspective of personal health risk and away from personal values and social reforms. Precisely because the story personalized medicine narrates about itself bears the Janus face of natural law and utilitarian pragmatism, we risk missing out on the potential for depersonalization.

8

Personalized medicine and genetic newborn screening

Margherita Brusa and Donna Dickenson

Introduction

Personalized medicine has always stressed the importance of knowing one's own genetic profile. The logical outcome of that view is that the earlier the profile is made available, the better—preferably from the very beginning.

From the early days, proponents of personalized medicine presented as a sort of Holy Grail the possibility that a baby could have its entire genome sequenced at birth (Collins, 2010; Collins. 2014; Angrist, 2010). Not only would whole-genome sequencing enable early detection of genetically linked diseases; more speculatively, it might also enable the child to benefit from made-to-order diagnostic tools, lifestyle recommendations, preventive measures, and drugs throughout her entire lifetime (Ginsburg and Huntington, 2009).

At the time, this proposal was more a statement of intent than a real possibility, but with the declining cost of whole-genome sequencing, it is increasingly feasible to sequence on a large scale. At its most far-reaching, the goal of newborn sequencing would presumably have to be universal, building on the precedent of other universal programmes available (or even required) for infants in many countries. Otherwise, inequality would be inscribed from birth between those newborns whose parents could afford the cost of sequencing and those whose families could not. Public health considerations might seem to dictate universal genomic sequencing, as much as they do universal blood tests for certain conditions, particularly if the benefits are conferred at no cost to the user.

On the other hand, universal adult screening for conditions such as breast cancer is already under attack for lack of specificity and sensitivity. Many such programmes have been reconsidered owing to a problematic harm:benefit

Margherita Brusa and Donna Dickenson, *Personalized medicine and genetic newborn screening* In: *Can precision medicine be personal; Can personalized medicine be precise?*. Edited by: Yechiel Michael Barilan, Margherita Brusa, and Aaron Ciechanover, Oxford University Press. © Oxford University Press 2022. DOI: 10.1093/oso/9780198863465.003.0008

ratio (Gøtzche, 2012; Ahn et al., 2014; Ito et al., 2019). The scattergun approach of newborn genetic sequencing might be expected to result in similar problems on a larger scale, transcending any single condition. Given that everyone carries a certain number of adverse genetic alleles, universal newborn sequencing might well result in a population of worried well or 'patients-in-waiting' (Timmermans and Buchbinder, 2010). It might create a reality in which every child is 'in waiting' for something, and every child is associated with a number of variants of uncertain significance (Friedman et al., 2020). Overall, every newborn will have a personalized genetic profile whose clinical significance is uncertain. 'Knowing thyself', in obedience to the maxim inscribed on the temple at Delphi, would become knowing one's own uncertainties, as well as one's risk profile relative to others.

In this chapter, we begin by examining the history and rationale of established medical enterprises of mass participation for newborns to draw lessons about genome sequencing at birth.

Newborn screening programmes: the background

Sixty years ago, an American physician invented two techniques that revolutionized public health. The first is soaking a few drops of blood into a piece of cardboard, allowing it to dry. The dried cardboards, also known as 'Guthrie cards', facilitate easy transport and long-term storage of a person's blood. The second invention was a laboratory procedure that detected in the blood specific metabolites typical of a rare genetic–metabolic disease: phenylketonuria (PKU). Finding that a newborn baby suffers from this metabolic disease allowed caregivers to nourish the child with a special diet that would save him or her from a devastating developmental disorder. In 1962, Massachusetts was the first state that initiated a pilot programme of screening of all newborn babies for PKU, with the intention of detecting all those affected before damage sets in. The PKU screening programme quickly disseminated all over the world. In the early 1970s, testing the dried blood spots was expanded so as to include early detection of hypothyroidism. There were attempts to add more screening tests to routine neonatal care, not all of which were by means of blood tests, but many such addition was labour-intensive and expensive.

In the 1990s, newborn screening (NBS) entered an era of significant expansion owing to the introduction of a novel technology for the analysis of the dried blood samples, called tandem mass spectrometry (acronym MS/

MS). MS/MS is very expensive, but, once acquired, it can process virtually endless biochemical tests on the Guthrie cards with no additional costs. Soon, many conditions were added to the screening list of diseases, also called screening panel, reaching up to sixty metabolic conditions, some of which are extremely rare and of unknown significance. Promoters of expansion argue that it is the only way to learn about the true nature of the most extreme metabolic 'disorders', with the hope of finding treatment. In the early years of MS/MS-driven NBS expansion (1990s–2000s), debates raged about which 'conditions' should be included in the panel, and which be considered 'pilot' or experimental. Additionally, with the conclusion of the Human Genome Project, many saw a future in which technology would be able to sequence and screen the genome stored in the Guthrie cards, much as MS/MS did in relation to metabolites.

The evolving discourse on genetic testing in general had a strong emphasis on individual counselling, informed consent, special data protection, and a right 'not to know'. The logic concerning NBS seems to have invited genetic testing as a universal standard of negligible ethical legal and social implications. Because the established NBS programmes already acquire a few blood spots from every neonate as a 'life-saving' measure, a transition to genetic screening does not even require separate sampling and, arguably, not even the informed consent process involved. Genomics technology and the unquestionable status of NBS have imbued genetic testing and storage with the logic of one of the most successful public health screening programmes.

In 1968, the World Health Organization (WHO) defined screening as the testing of a person only because he or she is in a high-risk group; all other medical tests are conducted following positive screening results, symptoms or signs related to the individual (Wilson and Jungner, 1968). Laboratories already rely on genetic technologies to establish diagnosis and resolve ambiguous screening results. NBS is a special case because the fact of being born is the only inclusion criterion for screening. Therefore, all babies are by definition at risk and deserve screening.

Today, many NBS programmes test for more than fifty conditions. These programmes have expanded from searching for a few catastrophic but preventable conditions to looking for early diagnoses of conditions whose nature is less clear and for which early intervention is quite limited. Advocates of expansion argue that mere knowledge of a future condition is beneficial to child and family. However, the leading motivation and success of NBS is rooted in those few diseases whose immediate postnatal diagnosis marks a huge and remediable difference for the child's health, life, and future.

Most of the conditions covered by NBS are genetic. Most known genetic diseases become manifest during childhood. Whereas MS/MS targets proteinomics, advances in genetics enable us to capture the genetic mutations responsible for the proteinomic abnormalities. The recent plummeting costs of genetic sequencing and the allure of personalized/precision medicine have prompted some scientific sectors to call for a third era of expansion—the application of genetic technology to NBS.

The incorporation of genomics into NBS involves diverse technologies. First, probes may be used to detect one or many mutations. Second, it is possible to sequence not only specific genes, but also greater parts of the genome, up to whole-genome sequencing, which will allow the detection of alterations that have not yet been described or that are so rare that no probe is available for them. Third, it is possible to create genetic profiles, which are statistical compilations by artificial intelligence of combinations of single nucleotide polymorphisms and other biomarkers. Because sequencing as such merely exposes and documents genetic variants,[1] algorithms are needed to interpret them.

Once a databank or repository of dried blood spots has been built up, researchers could theoretically use it to expand the horizons of NBS by also looking for all Mendelian conditions. Additionally, they could search for the relative risks of the baby developing numerous future health problems (e.g. obesity, cancer, and degenerative diseases) or, perhaps, they could even try to find correlations that are non-health-related such as intelligence or psychological traits, although these uses would be speculative and controversial (Plomin, 2018).

NBS has developed as an impersonal and universal practice, whose aim is the prevention of a few rare medical catastrophes. It is impersonal and universal because all newborns are subjected to the very same panel, regardless of any personal signs and symptoms. Most infants (>95%) return negative results in all screening tests. About this vast majority, NBS has revealed nothing. The expansion of NBS to whole-genome sequencing will inevitably tell us something about everybody. Genetic analysis and the application of sophisticated algorithms may turn NBS into an instrument that scrutinizes everybody's genome, extracting from it a unique profile that differentiates the baby from all others, ideally with the intention of tailoring best medical advice either in the near or distant future. While 'traditional'

---

[1] About 99.8% of the genome is identical in all people.

screening would fish out only those individuals in need of urgent care, genetic screening would greatly widen the horizon and scope of the programme, rendering every single baby a prospective patient.

The bioethicist George Annas described such personal information as a 'probabilistic future diary' written in a code that has not yet been broken (Annas, 2000). Twenty years later, it seems that 'big data' algorithms may decipher the diary in terms of relative risks and chances (Topol, 2019, pp. 205–15).

Genetic testing of babies

Currently, genetic analysis of babies is indication-based only. The indication might be a clinical suspicion, an undiagnosed serious illness, or an aberrant newborn screening result. Some NBS services assume exemption from informed consent to the genetic component of testing because the disease is already in the authorized NBS panel.[2] However, such practice is limited to targeting loci associated with the condition in question, while ignoring all other genetic information and material (King and Smith, 2019). Except for acutely ill babies, for whom urgent genetic analysis might confer substantial clinical benefit, genetic counselling and informed consent to genetic testing are considered mandatory (Gyngell et al., 2019). It is noteworthy that only a tiny fraction of newborns and children are candidates for genome sequencing because the clinical data offer no clue to the diagnosis of the serious conditions from which they suffer.[3]

The addition of Mendelian conditions to NBS constructs 'genetic information' as pairing specific mutations with highly penetrant conditions. Because every added screening item is comprised of a genetic variant paired with an International Classification of Diseases diagnosis, a clinical trial (pilot

---

[2] Every NBS programme screens for more than one condition. The list of conditions, which varies from country to country, is called a 'panel'. See, for example Genetic Alliance UK 2019, p. 11, Figure 1.

[3] Around 10% of newborns are hospitalized in newborn intensive care units. The vast majority suffer from non-genetic problems such as prematurity. Most of those affected by genetic diseases are diagnosed by clinical criteria. Genetic analysis confirms the suspicion of a disease or syndrome. In one retrospective study, out of 2400 babies in intensive care, 35 met the criteria of undiagnosed condition. Twenty were diagnosed by rapid genome testing of all known Mendelian disorders (n = 4300), leading to treatment modification in eight children (Willing et al., 2015). In another study, similar rapid sequencing led to a rapid diagnosis in 30% (7 out of 23) of critically ill children for whom the routine clinical work-up was unsuccessful (van Diemen, 2017).

study) may be offered to parents. The UK National Screening Committee (2015) requires evidence based on 'high quality' randomized clinical trials as a condition for incorporation of a new screening test in the panel. Many screening programmes have expanded following much laxer criteria (Botkin et al., 2014).

The BabySeq project is the most recent attempt to examine the value of expanding screening to all Mendelian conditions. The researchers 'curated' from the literature the most comprehensive list of gene–disease pairs and, in the framework of a clinical trial, tested 159 newborn babies, healthy and sick alike, for 954 conditions that met a complex set of criteria, which combines considerations of quality of evidence, degree of penetrance, and actionability. The curation process did not refer to variations in expressivity. Because it aimed at many mutations that are not ordinarily tested, the BabySeq project needed to use rapid sequencing techniques (next-generation sequencing).

This technique identified a risk for childhood-onset disease in 9.4% of babies, and risk for actionable adult-onset disease in 3.5%. Genes related to cardiomyopathies were detected in 4% of the babies, a rate much higher than expected. None of these babies had a suggestive family history or other indication for suspecting the condition. It is noteworthy that some genes, previously reported as important causes of a given cardiomyopathy, have been found to be a rare variation. Hence, they are not clinically informative (Walsh et al., 2017). Other positive findings are related to the most common genetic disorder (G6PD deficiency), which is discovered following a haemolytic crisis and is not considered life-threatening. Of these, 3.5% were positive for genes associated with adult-onset disease, mainly cancer-related genes.

BabySeq is an example of a clinical trial that required institutional review board approval and full, documented informed consent. The research protocol was quite thorough in its coverage of different result scenarios and in its commitment to proper counselling, together with long-term follow-up of babies. The BabySeq project did not act on the assumption that 'genetic knowledge' is principally beneficial and almost always innocuous. Even 'information worth returning' by the project's criteria needed individual counselling. Moreover, the project team concludes that many novel variations require laborious laboratory work in order to become meaningful information (Ceyhan-Birsoy, 2019, p. 88).

Screening for Mendelian and other discrete conditions tells us nothing about the future health of the child who is screened negative. Indeed, the mass screening of all babies is relevant only for the hidden 0.1% who will benefit substantially. The phenotype-based structure of NBS preserves the paradigm of public health by being disease-centred, committed to strong scientific evidence and being strictly limited to dealing with specifically defined targets, be they nine (the current UK screening panel) or 56 (the panel recommended by the American College of Obstetricians and Gynecologists). Clinical research and ethical deliberation may inform us whether specific genetic targets should be included, such as some genes associated with cardiomyopathy, or excluded, such as genes related to late-onset disease.

The conclusion is that when NBS incorporates additional Mendelian conditions, it does not fit the paradigm of 'personalized medicine' because, even if it entails a certain medicalization of birth, the screening will be strictly limited to a closed list of rare and very rare conditions. Moreover, only positive screening results and clinical symptoms and signs will invite further testing, genetic or otherwise. Finally, NBS will remain a public health operation by keeping its characteristics of being uniformly universal and mainly preventive.

UK policy papers as a test case

Even at the beginning of the 2000s, well before the concept of 'personalized/precision medicine' had taken root, the vision of rendering medical care more personalized and more precise by means of genetic analysis appears in official documents. Attempts to introduce the idea of genetic newborn screening surfaced in the UK National Health Service (NHS). One of the first documents that states the possibility of screening the entire genome of newborn babies is the 2003 NHS White Paper on Genetics, titled 'Our Inheritance, Our Future: Realizing the Potential of Genetics'. In the subsection titled 'Genetic profiling at birth', it states:

One long term possibility that has been suggested is to screen babies at birth as part of the standard postnatal checks and to produce a comprehensive map of their key genetic markers, or even their entire genome. [3.36, pp. 44–5]

The baby's genetic information could be securely stored on their electronic patient record for future use. It could then be used throughout their lifetime to tailor prevention and treatment regimes to their needs as future knowledge becomes available about how our genes affect our risk of disease and our response to medicine. [3.36, p. 45]

The fact that these suggestions appear in the UK is quite surprising and interesting. In fact, while pioneering the idea of applying genetic analysis in newborn screening, the UK is known for being quite conservative in relation to the number of diseases listed in its NBS panel, strictly following the WHO criteria of screening (Wilson and Jungner 1968; UK National Screening Committee, 2015). As in other developed countries, such as France, the NHS screens babies for only nine diseases, whereas others, the USA for example, screen for over fifty conditions. The gap between the strong evidence-based medicine criteria used for the inclusion of NBS testing feels at odds with the actual tendency towards widening newborn screening in the direction of universal genomic sequencing of newborns, with the stated goal of building genetics into mainstream services (Ross et al., 2013).

Following the Government's strategy for maximizing the potential of genetics in the National Health Service, and being aware of technical problems as well as legal and ethical issues related to this goal, the Human Genetics Commission and the UK National Screening Committee formed a joint working group to explore prospective gene technology for the newborn. The results of this joint effort were published in the 2005 report, titled 'Profiling the Newborn' (Human Genetic Commission, 2005).

The report underscores the difference between *screening* and *profiling*. Screening aims at a specific disease or diseases within a population at risk. Profiling aims at revealing variations that characterize personal genetic information. Therefore, genetic profiling does not expand newborn screening but marks a 'radical shift in that it would identify the majority of genetic variants as opposed to variants related to a specific disease' (Human Genetic Commission, 2005, p. 10). Moreover, since the link between most genetic variants and a disease is not clear, more knowledge is necessary to interpret the significance of the data collected. The collection of variants of unclear significance cannot be the basis of a screening programme.

Promoters of personalized medicine consider genetic NBS to be a logical extension of NBS (Kingsmore et al., 2015). Indeed, according to one rationale of personalized medicine, there is no need for indication-based testing. Instead of waiting for people to show up with symptoms and then gathering information about their 'condition', medical practitioners will already have much of the necessary information stored from birth, ready for deciphering whenever there is something to know or to do about impending illnesses. Instead of focusing on very rare cases and pathologies, for which genetic testing in some babies and children may well be justified, they present genetic testing as a kind of preventive tool for future diseases, a repository of data ready to use when necessary. To make this move, they wish to benefit from the credibility and success of NBS programmes and to represent its proteinomics and genomic expansions as a logical consequence of these developments. 'The rationale for taking a sample at birth is that it would be relatively convenient to collect, and that the linkage of such data to a person's medical record and care might confer a clinical advantage' (Human Genetics Commission 2005, p. 11).

It follows from this report that the only context justifying profiling of the newborn is clinical research. Even though blood spots are taken routinely, using them for profiling would be qualitatively different. The screening use of the spots is a well-established public health service (which nevertheless depends on parental consent in most jurisdictions). By contrast, profiling the very same blood spots constitutes genetic research on minors that has no clear benefit yet to any group of children. Evidently, such a project is driven by a general presumption of mining novel and useful clinical information. Yet, this hope is too broad and too nebulous to stand shoulder to shoulder with screening for specific diseases.

In 2011, the UK Government abolished the Human Genetics Commission (Hill, 2012). In its last report, the Commission wrote recommendations to the departmental expert committee Emerging Science and Bioethics Advisory Committee (ESBAC). Until it came to an end in 2014, this new committee published neither on NBS nor on genetic testing on minors.

In the *Annual Report of the Chief Medical Officer 2016: Generation Genome*,[4] there is a chapter on genomics in newborn screening, in which the author warns, "[T]he implementation of NGS in NBS should not be technology-driven and should always be for the health interests of the child" (ch 11, p. 8). The report reiterates the conclusions expressed by the Human Genetic Commission. The Chief Medical Officer endorses the use of next-generation sequencing in NBS only when it is focused on variants of high penetrance, for which there are effective and accepted therapeutic interventions, needed during early childhood.

In 2019, Genetic Alliance, a UK charity concerned with rare diseases, published a draft report titled *Fixing the Present: Building for the Future*. The report, at p. 18, states that NBS is committed to assisting sufferers of rare diseases, endorsing the US Newborn Screening Expert Group's statement that 'newborn screening policy development should be driven primarily by the

---

[4] https://assets.publishing.service.gov.uk/government/uploads/system/uploads/attach-ment_data/file/631043/CMO_annual_report_generation_genome.pdf

interests of affected newborns, with secondary consideration being given to the interests of unaffected newborns, families, health professionals and the public' (Watson, 2006). From this perspective, the report states that '[T]he UK National Screening Committee has failed to adapt its policy and methodology to treat rare diseases fairly, leaving the UK out of step with the majority of high-income countries' (Genetic Alliance UK, 2019, p. 26). Even though the report does not present any new conditions whose discovery upon birth is expected to benefit the child, the report calls for thorough revisions (the 'fixing') of the inclusion criteria to NBS, calling for 'a vast increase' in the number of rare diseases screened.

The report does not make this statement particularly in the context of genetic screening and testing. Its ethical claim to grant higher importance to sufferers from rare diseases relative to the consideration of all the rest is contrary to Article 8 of the International Declaration on Human Genetic Data: 'In diagnosis and healthcare, genetic screening and testing of minors and adults not able to consent will normally only be ethically acceptable when they have important implications for the health of the person and have regard to his or her best interest' (UNESCO, 2003).

This apparent gap between scientific evidence and the proposed revision of practice is bridged by a novel conceptualization of 'expansion'. It is a move from screening for targeted panels of diseases to collection from the newborn of genetic data as such. The report calls for the establishment of a genetic repository of genomic information, an unprecedented goal in any screening programme. Present and future processing of the data in the repository will allow medical science to come up with the evidence now lacking for adding the vast number of additional conditions. Hence, the report advocates screening without evidence, on the assumption that the evidence will appear later.

It follows that research on pooled and unrestricted genetic data, possibly up to whole- genome sequencing collected from the newborn, is represented as an 'urgent duty' of fairness towards children with rare diseases (Ross et al., 2013). Whereas many experts regard the unique genetic profile of a person as a private secret diary (Markett, 1996; Everett, 2004), others believe that sharing genetic data is a moral duty (Chadwick and Berg, 2001; Mittlestadt et al., 2018), even in relation to children (Hens et al., 2011).

Promoters of personalized medicine believe that profiling the 'diaries' is necessary for benefiting everybody. Paradoxically, public health underscores the personal status of the genome, while promoters of personalized medicine call for universal sequencing as key to unleashing the hidden powers of preventive and curative medicine. This rhetoric of urgency and personalized medicine's promises loosens the methodological and ethical boundaries between evidence-based practice and basic science research.

Consent, solidarity, and sequencing the newborn

In November 2019, the UK Secretary of State for Health announced the intention to establish a pilot study sequencing the genomes of newborn infants. A few days later, the executive committee of the British Society for Genetic Medicine published its objection against sequencing babies without clinical justification, endorsing the prevailing screening criteria set by the NSC.[5] The Secretary of State for Health described genetic screening as 'life-saving'. However, an opinion expressed in the *British Medical Journal* referred to this promise as 'extremely misleading'. David Curtis, honorary professor at University College London's Genetic Institute, affirmed: 'It's difficult to think of any circumstances in which such a test could make a substantial difference, with the exception of tests for cancer-causing mutations such as in the BRCA genes. But if these are indicated they can readily be performed in an adult who can consent to the testing procedure'. Professor Curtis concluded that '[T]he only justification to perform an investigation on a child who is unable to provide consent is that some immediate action is required' (Mahase, 2019).

If, in the great majority of cases, there will be no clinical justification for screening newborns, then there is no firm legal basis on which parental consent should or can be sought. That statement sounds extreme and surprising, but in the common law, which governs the UK proposals, parents give informed consent on behalf of their newborns in the name of the child's best medical interest. Generally, the law assumes that parents have their child's best medical interest in mind, although in rare cases, courts have sometimes intervened to protect the child's best medical interests against the parents' decision (UK Supreme Court, 2017). The standard remains the child's best medical interest, however; parents have no independent right to consent on behalf of the child, except to serve that standard.

There is an alternative justification for sequencing newborns: that like the construction of national biobanks such as UK Biobank or the Icelandic database, it purportedly serves the common good, represents altruism, and promotes solidarity (Brusa and Barilan, 2017). Indeed, the Secretary of State for Health's original statement spoke about newborns as being 'volunteered by their parents', implying the same motives of altruism commonly attributed to participants in research trials. Legally, this argument is suspect, however. If parents can only act in the name of the child's best medical interests, they have no right to 'volunteer' their newborns even in the name of altruism or solidarity. We cannot presume altruism in newborns, of course, but neither can we assume that the adults those newborns will become would have wanted to 'volunteer'.

If screening and sequencing uncover risks that can be remedied, then the child's best interest does seem to be served. Even in that case, however, we need to recognize the difficulties of engaging parents in an appropriate informed consent process, tackling issues such as how to explain each of the possible conditions targeted, how to understand unclear or probabilistic results, how to interpret incidental findings, and whether to report carrier status. Caution is called for, but unfortunately the dynamic seems to us to be moving in the opposite direction, towards lessening or even bypassing informed consent in the case of adults, despite the common law's traditional presumption of their competence and the right to refuse consent to treatment or research (Lunshof et al., 2008). *A fortiori*, children, who are presumed to lack capacity, are at risk of having their consent presumed, if their parents ostensibly give it on their behalf.

In 2014, the US Newborn Screening Translational Research Network published guidelines on pilot research in NBS (Botkin et al., 2014). The guidelines assume that, in the context of large-scale clinical trials (called 'pilots'), full written informed consent may be impracticable, risks to infants are minimal, and no harm should come to the rights and welfare of the subjects. Therefore, they claim that 'a waiver of a formal permission process with a signed consent document is appropriate in some circumstances'.

NBS is oriented to a definite set of clinical questions—'does the baby have this or that condition?' NBS databases are a by-product of collecting samples from numerous babies. Advocates of whole-genome sequencing and profiling of the newborn reverse this order, seeking first to create a databank and then to see whether useful clinical information might follow. But the distinction between a database as a by-product of clinical activity and a databank is crucial. One of the first definitions of databanks was a 'generalized collection of data not linked to one set of functional questions' (Malik, 1971). It follows that even though genetic sequencing of the newborn might look similar to NBS, the combination of complete genetic exposure lacking a functional (i.e. actionable) clinical question renders genetic sequencing and profiling of the newborn morally problematic. In the absence of concrete hypotheses, it can hardly fall within the category of clinical experimentation on humans. Instead of science searching for a procedure that can prevent illness, whole-genome sequencing is a procedure in search of illnesses. In this light, as an unprecedently broad exposure of a child with no clinical indication or any specific question, obtaining parental consent and reporting 'findings' become a new challenge (Johnston et al., 2018, S22; Biesecker and Green, 2014).

There is an inherent contradiction between personalized medicine's catering to 'me medicine', the benefit to the unique individual, and the value of solidarity, which is about a choice to make some sacrifices for the sake of others. It would be absurd to find personalized medicine promoting such a personal act as whole-genome sequencing without personal informed consent. Moreover, the consent needed might be too broad in scope and, consequently, under-informed. If there is no individual information involved, the issue is much less pressing; yet, with such a comprehensive exposure (whole-genome) and comparative scales (profiling), there must be some 'information' that somebody might deem relevant. Nobody knows what kind—perhaps a risk for cancer in old age, a psychiatric disease in middle age, or juvenile diabetes. The scope of whole-genome sequencing is always vastly wider than all preventable medical harms put together. Hence, sequencing always produces personal data whose meaning nobody yet understands. This exposure of personal information of unclear significance poses a unique challenge to the values of informed consent and the best interest of the child.

Financial conflicts of interest

We have argued that even with an expanded list of conditions being tested, newborn screening in anything like its current form does not readily match the model of personalized medicine. Although other population-level forms of screening promote solidarity and public health, these are not the primary motivations behind personalized medicine's call for individual newborn sequencing, which is much more about 'me' than 'we' (Dickenson, 2013). As opposed to public health operations, including NBS where it is usually offered as a publicly funded service, personalized medicine embodies the age of private entrepreneurship in healthcare.

Extending the model of personalized medicine to whole-genome sequencing at birth does seem to conform to personalized medicine's inner logic and expansionary drive. However, would the alleged benefits justify what would clearly be a huge financial expansion and investment?—not only in delivering and interpreting the sequencing, but in follow-up counselling and possible interventions. Newborn genetic sequencing would ride on the 'urgency narrative' behind other forms of NBS (Grob, 2019); but that, we feel, would be undeserved.

In healthcare systems that do not provide free or partially subsidized medical cover for their entire population, universal mandatory testing for a limited range of conditions can be an effective means of reaching vulnerable populations who would otherwise slip through the net (Johnston et al., 2018). These services often operate within tight limits of budgets and staff; expanding their remit to universal whole-genome sequencing for newborns would almost certainly be impossible. Given the plethora of commercial providers in the direct-to-consumer genetic testing market, however, it is entirely feasible that governments might outsource the testing programmes to private providers. The business strategy of those providers might well be the accumulation of large databases—so large as to include all the newborns in the country—for purposes of patenting and contracts with pharmaceutical firms (Dickenson, 2013, ch. 7). Such a development would be in line with social media's business model, in which a simple service is distributed for free, while turning people's sensitive data into a commodity that is sold for profit.

We need to think long and hard about whether this extensive expansion and involvement of commercial interests is what we want to see for our newborns and our health services.

Conclusion

In this chapter, we have asked whether personalized medicine can justify genetic sequencing of newborn babies. We have examined whether the attempts to introduce genetic-based healthcare into public health services would justify the use of the already well-established NBS programme as a basis for profiling each individual baby's genome, thereby revealing all of his or her genetic variations, even before their clinical significance (if any) is known.

We have found several possible meanings for 'genetic expansion' of NBS (Johnston et al., 2018). The first is the inclusion of many more—up to all—Mendelian conditions. This expansion targets pre-defined medical conditions. The second is genome sequencing so as to allow profiling, which exposes the variations among individuals in terms of risk. The third is the creation of a repository of genetic data, to be processed and stored for future use by the child or present use by databank managers, public or private. The fourth is the merging of NBS with personalized medicine's collection of genetic, proteinomic, and other forms of data in search of optimizing people's wellness.

We have found that all these kinds of expansion amount to genetic research on minors, since there is no evidence-based medicine about the practice nor professional consensus that supports it. Some forms of expansion—profiling and sequencing—can hardly count as medical research because they operate without specific clinical questions and ends.

Inclusion of additional Mendelian conditions and any novel screening test aimed at actionable childhood disease must follow that track of medical experiments on minors. Mere sequencing, profiling and data banking do not qualify as 'research'. Hence, we do not see how minors might be subjected to these practices even if parental consent is granted. Moreover, public health services should not solicit parental consent for interventions that are not to the clear benefit of the child.

Solidarity is a leading justification to research on humans, but not a justification to deviate from the ethical standards of this research. One of these standards is the avoidance of research on minors that can be done when they come of age. This is most relevant in the context of personalized medicine, in which personal choice and participation are key, especially in relation to self-knowledge and self-monitoring. It is paradoxical that public health underscores the personal status of the genome, while promoters of personalized medicine call for universal sequencing as key to unleashing the hidden powers of preventive and curative medicine.

Is it necessarily in the baby's best interest to have her genome sequenced if there is no reason to suspect a genetic condition? To turn the argument around, would people want to have had their genomes sequenced when they were children? Many might feel that they could be forced to live with information that they do not want, a particular concern given the imperfect, complicated, and often erroneous state of genomic information at present. The child's right to an open future is often seen as the correct underlying principle, except in cases of early-onset disorders or difficult-to-diagnose cases in newborns (Worthey et al., 2011). The oncologist Siddhartha Mukherjee has drawn attention to the sense of 'being under siege from the future' (Mukharjee, 2017). In our natural and laudable desire to do the best for our children, through universal whole-genome sequencing at birth, we should not unwittingly encumber them with that burden.

The urgency narrative on which the success of the traditional NBS programme has been built all over the world cannot be applied to the performance of genetic screening on all babies. There is a huge distance between preventing a catastrophic disease in a baby (such as PKU), where timely and effective intervention is needed, and the collection of precious personal data of unknown significance for no immediate clinical benefit.

Whole-genome screening may or may not be the scientific pathway leading to the future of medicine; but, like any other medical promise, it has to prove itself through basic and clinical research. It should not infiltrate clinical standard childcare and overturn legal presumptions around consent in the guise of an expansion of a public health initiative for children.

9

The revolution of personalized medicine is already upon us in rare diseases

Christopher P. Austin

Introduction

Personalized medicine remains an aspirational *future state* for the majority of patients and physicians, since they are affected by or treat common diseases such as heart disease, diabetes, and Alzheimer disease. But personalized medicine is the *present state* for the field of medicine that diagnoses and treats people living with rare genetic diseases. Among medical fields, none has been affected more dramatically and positively by the technologies that enable personalized medicine than the field of rare diseases. In fact, though it is remarkably underappreciated, it can be confidently asserted that the principles, practices, potential, and perils of personalized medicine are already instantiated in rare disease diagnosis and treatment. In many ways, we can look through the lens of rare disease present and see the future for common disease personalized medicine.

The definition of a 'rare' (sometimes also called 'orphan') disease varies somewhat among countries, but all approximate a population prevalence of less than one in 2,000. In the USA, the designation of rare diseases as affecting fewer than 200,000 was codified in law by the Orphan Drug Act of 1983. Other countries soon followed with their own laws or regulations, including Japan (1993), Australia (1997), and the European Union (2000). A notable feature of virtually all country and region designations has been their motivation by political activism of rare disease patients and their families, who wished to create incentives for research and drug development to 'level the playing field' with more common diseases (Institute of Medicine, 2010).

There are currently ~7000 different rare diseases defined, and ~250 new rare diseases are being defined every year with advances in genomic sequencing and clinical phenotyping. At the extreme of rarity are the $N = 1$ diseases

Christopher P. Austin, *The revolution of personalized medicine is already upon us in rare diseases* In: *Can precision medicine be personal; Can personalized medicine be precise?*. Edited by: Yechiel Michael Barilan, Margherita Brusa, and Aaron Ciechanover, Oxford University Press. © Oxford University Press 2022.
DOI: 10.1093/oso/9780198863465.003.0009

that are being discovered through programmes such as the Undiagnosed Disease Network International (UDNI, https://www.udninternational.org/); about 30% of patients with rare syndromes are undiagnosable with any current technology. Though each rare disease is by definition uncommon, given their multiplicity they are cumulatively quite common: population prevalence is estimated at 25 million people in the USA, 30 million people in Europe, and 400 million people worldwide—about the same as diabetes (Posada de la Paz et al., 2017).

Only about 400 of the more than 7000 rare diseases have a specific treatment approved by a regulatory agency such the US Food and Drug Administration (FDA), European Medicines Agency (EMA), or Japan Pharmaceuticals and Medical Devices Agency. The majority of rare diseases have their onset in childhood, and many (though certainly not all) are characterized by high degrees of morbidity and premature mortality. Most (~80%) are caused by a genetic lesion of large effect, and demonstrate Mendelian inheritance; familiar examples are sickle cell disease, Tay–Sachs disease, cystic fibrosis, and Huntington's disease.

Rare disease medicine

Rare disease medicine—that is, the practice of medicine that focuses on the care of individuals with rare diseases—is operationally and culturally distinct from traditional mainstream medicine. There is no recognized specialty of 'rare disease medicine'; rather, its practitioners are generally trained in medical genetics as well as a particular clinical specialty such as paediatrics or neurology. Whereas doctors are trained to think of common diagnoses first when confronted with symptoms—the adage 'when you hear hoofbeats, it's probably a horse, not a zebra' is taught widely—rare disease doctors are trained to think of uncommon diagnoses first, and in fact the zebra is the unofficial mascot of rare disease patients and doctors. This difference manifests in what is known as the rare disease 'diagnostic odyssey': the migration of patients from one doctor to another over 5–15 years without a correct diagnosis, which ends when the patient eventually sees a rare disease physician trained in 'zebra' diagnoses. In personalized medicine for common diseases, this same dynamic will likely be at play.

Because the majority of rare diseases are caused by inherited mutations in single genes, their diagnosis is now largely achieved by genetic testing, either of individual genes or of the entire exome or genome. Most (but certainly not all) rare genetic diseases can be caused by many distinct mutations in the same gene, each of which can cause a somewhat different clinical presentation and dictate different prognoses and treatment strategies. For example, there are more than 600 different mutations in the *dystrophin* gene that cause Duchenne muscular dystrophy, but some of those mutations result in more severe dystrophin protein dysfunction than others, and symptom progression and life expectancy track with that severity. As rare disease diagnoses have shifted from being symptom-based to gene-based, it has been recognized that a single symptom-based clinical syndrome (or 'disease') can be caused by mutations in multiple different genes. For example, the inherited progressive blindness disorder diagnosed clinically as retinitis pigmentosa is now known to be caused by mutations in over 50 different genes. Conversely, completely different clinical syndromes can be caused by different types of mutations in the same gene; for example, mutations in *LMNA*, which codes for the protein lamin A, can cause disorders as disparate as Charcot–Marie–Tooth peripheral neuropathy, dilated cardiomyopathy, and Hutchinson–Gilford–Progeria premature ageing syndrome.

This 'personalized medicine' approach has led directly to more accurate diagnoses, improved clinical management, and new treatments for many rare diseases. But the term 'personalized medicine' is not used in the rare disease community, since it would be redundant. Rather, 'personalized medicine' is generally applied to common diseases, where it refers to technology-enabled segmentation of common illnesses from the general to the specific, and often the syndromic to the molecular or genetic. The department of the US National Institutes of Health that oversaw the Human Genome Project defines personalized medicine as 'an emerging practice of medicine that uses an individual's genetic profile to guide decisions made in regard to the prevention, diagnosis, and treatment of disease' (https://www.genome.gov/). This definition is aspirational when applied to common diseases but it is the current reality for rare diseases. And as currently common (in the USA >200,000 prevalence) diseases become increasingly segmented based on molecular characteristics, those diseases are transitioning to being collections of clinically similar but molecularly distinct rare (in the USA <200,000 prevalence) diseases. This transition is most advanced in certain types of cancer, but it is possible to imagine a time, decades in the future, when all diseases will be 'rare'.

Implications for clinical management and treatment development

Personalized medicine endeavours to use in-depth genetic and environmental data to define a genotype–phenotype pairing that is based in definable physiological dysfunction, and it predicts treatment with a high likelihood of positive response. Rare diseases are defined and managed in just this way today. When successful, the 'rare disease/personalized medicine' paradigm is remarkably effective. Because rare diseases are frequently metabolic or Mendelian genetic, advances in analytical chemistry and genome sequencing have revolutionized their diagnosis. And treatments can be truly transformational—as demonstrated by the treatment of children for phenylketonuria, Gaucher disease, cystic fibrosis, metastatic melanoma, and spinal muscular atrophy—allowing diseases that had been universally and often rapidly fatal to be stopped, reversed, and even cured. The same is true for some common diseases that have been segmented into biologically more uniform subsets with higher response rates to targeted therapies, HER2-positive breast cancer and ALK-positive lung cancer being two prominent examples.

However, the current financial dynamics of drug development often result in these highly effective rare disease therapies being extremely expensive. Since the costs of developing a new drug are roughly the same regardless of the prevalence of the disease it is intended to treat, the relatively fixed costs of development must be recouped by use in use in a smaller number of patients in the rare/personalized medicine scenario. This results in drug prices that are inversely proportional to disease prevalence, in order to maintain return on investment for their developers (Figure 9.1).

Rare disease drugs approved in the last decade typically carry a list price of between US$300,000 and 600,000 per year, and the treatments are life-long. Recently approved gene therapies carry a price of US$1–2 million per patient, though they are thought to be one-time curative therapies. These prices are 10–100 times those of previous therapies. Most health authorities worldwide pay for these drugs, since they are effective and the numbers of patients needing them has so far been small. But the potential return on investment has attracted increasing numbers of companies to pursue rare/personalized therapies, and so their number is increasing rapidly: approximately half of all drug approvals by the FDA and EMA every year are now for rare (or 'orphan') conditions. The increased success rates and high prices commanded by these treatments has contributed to decreases in drug development for major causes of morbidity and mortality on a population level, including

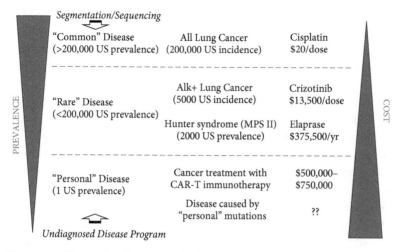

Figure 9.1 'Personalized medicine' is obverse of 'rare disease medicine' (prices shown for US market).

Alzheimer's disease, bipolar illness and schizophrenia, chronic kidney failure, and drug-resistant infectious diseases. Arguments for health equity for those living with rare diseases have furthered these developments, particularly in Europe (Szegedi et al., 2018). Virtually all countries worldwide operate within fixed health budgets that are being put under strain by the increasing availability and high prices of rare/personalized medicine therapies (Ollendorf et al., 2018).

Patient advocacy in rare diseases

Patient advocacy is a prominent feature of the rare diseases landscape and in many ways has shaped research, funding, care, and policy. Since most rare diseases have onset in childhood or are severely disabling, most advocates are parents. In the USA, the mother of a boy with Tourette's syndrome, Abby Meyers, was a major driver of the creation of the Orphan Drug Act, and went on to found the National Organization for Rare Diseases, the largest rare disease patient advocacy group in the USA. Thousands of other parents and patients have now done the same, starting foundations and more recently moving from political advocacy and communication for patient care into driving and funding research and development of drugs for their own diseases. The stories of John Crowley and Amicus, Kathy Giusti and the Multiple Myeloma Research Foundation, Pat Furlong and Patient Project Muscular Dystrophy, and Karen Pignet-Aiach and Lysogene are exemplary of the direct personal involvement of patients which has so benefited rare disease research, care, and policy (Anand, 2009).

Another striking characteristic of the rare disease community is the extent and enthusiasm of international cooperation and collaboration. Since rare diseases are by definition rare, sometimes affecting as few as one in a million people, most countries have only a few patients with any rare disease, and researchers and knowledgeable practitioners are equally uncommon. Aided by modern telecommunications and especially the Internet, rare disease patients, researchers, and physicians have formed international partnerships to actively share information and collaborate on research, including clinical trials (Walkley et al., 2016). This has reached its apotheosis in organizations like the International Rare Diseases Research Consortium (https://irdirc.org/), with over 50 funder, researcher, industry, and patient members from over 20 countries on five continents, and Rare Diseases International (https://www.rarediseasesinternational.org/) a network of over 70 member organizations representing rare disease patient groups in over 100 countries worldwide.

Implications for personalized medicine

The experience of rare diseases suggests both exhilarating medical promise, and sobering operational and ethical realities, for personalized medicine of common diseases. Rare disease/personalized medicine can be just as effective as anticipated. But current medical systems, drug development, treatment, and payment paradigms are not well-suited to personalized medicine, and all will require adaptations for its promise to be realized.

Sustainable personalized medicine will require scientific innovations, such as individualized tissue chips derived from induced pluripotent stem cells (Low et al., 2020). New personalized somatic gene editing and gene therapy technologies, which are directed at disease-causing mutations that may be unique to that individual ('personal mutations'), raise additional challenges since they may only be useful for only a single person, and may only need to be administered once to provide lasting improvement, even cure. These technologies are being developed and applied at a pace that is unprecedented in the history of medicine, including potential application to heritable germline gene editing. Although progress is to be celebrated, the technologies may be applied in inappropriate or potentially dangerous ways; ethics guidance is struggling to keep up (National Academy of Medicine, 2020). To address the cost issues, scientific and operational innovations in translational science (Austin, 2018) are needed to decrease the cost of production of new therapeutics, and implement new operational models that can produce treatments for small populations at manageable cost, and balance the resources being applied to personalized medicine, large-population medicine for common diseases, and public health interventions.

Concern has been raised that personalized medicine might compromise the patient–doctor (or more broadly, patient–healthcare professional) relationship, which can be therapeutic in its own right (Smith and Thompson, 1993). Experience in rare diseases provides reassurance that, properly managed, this relationship need not be 'depersonalized' by personalized medicine, but rather may be enhanced by it. The precise diagnoses and prognoses that characterize personalized medicine can provide relief from uncertainty in patients and caregivers, and often have implications for care even if there is no specific regulatorily approved targeted treatment. When targeted treatments do exist, they often have large clinical and functional effects on the patient's health and wellbeing of the sort previously only seen with antibiotics for bacterial infections. The dramatic effects of genotype-driven therapies for cystic fibrosis and spinal muscular atrophy demonstrate this promise, transforming previously universally fatal diseases into chronic, manageable disorders.

Rare diseases are frequently characterized by deep involvement of patients/family members as partners not only in care, but also in research, diagnosis, and treatment development, increasing their relevance and efficiency, and improving trial recruitment and adherence to protocols and treatment regimens. Perhaps paradoxically, rare disease research has led to unprecedented community efforts to share data, establish interoperable registries, and identify commonalities among diseases that would benefit all; this community-mindedness can be transformative to research and care, and promises to do the same for personalized medicine applied to common diseases.

At the same time, the high cost of treatment can fracture relationships between patients who may have different access to care in wealthy countries, and can be entirely out of reach for the majority of patients in less developed countries with small health budgets (Szegedi et al., 2018). The ongoing development of ex-vivo gene therapy for sickle cell disease is a notable example of a treatment that will be cost-prohibitive to the majority of affected individuals globally (Ledford, 2019). Further, involvement of patients/family members as partners in research and care can lead to counterproductive misunderstandings and slow progress, so this need to be managed carefully, which can be challenging in the context of time-limited clinical encounters.

On an operational level, personalized medicine is frequently antithetical to current systems of care, which are tailored to detect and treat highly prevalent diseases in large populations. This system has served society well, as infectious and cardiovascular diseases effectively addressed with the 'one size fits all' paradigm were conquered, resulting in a doubling of life expectancy during the twentieth century. But this paradigm assumes that (for the purposes of the intervention) all people are essentially the same; personalized medicine assumes the opposite—that all people are essentially different. This personalized medicine paradigm has begun to transform cancer care (Schilsky, 2014), and will need to be applied to other diseases as the science continues to evolve.

Finally, in our social media savvy era, the ethically fraught choices presented by personalized medicine are made even more complicated by the defining feature of personalized medicine, which is that it is *personalized*. Population health is anonymous, allowing dispassionate discussion based on hypothetical persons with broad public health benefits. Personalized medicine, instantiated today in rare disease medicine, is also media-friendly medicine, particularly since rare diseases disproportionately affect children. Rational discourse on ethical allocation of scarce resources or balancing the need of one ill person with the needs (or rights) of other ill persons easily gives way to social media storms that prioritize damage control over thoughtful resolution (Mackey and Schoenfeld, 2016). This is a manifestation of laudable and desirable care of individual human beings for other individual human beings. But it creates moral and ethical dilemmas that are likely to be increasingly common in the era of personalized medicine.

10

Personalized medicine and disorders of consciousness

An alternate convergence of knowledge towards a new clinical nosology

Joseph J. Fins

Introduction

When scholars write about personalized medicine, most refer to the interpretation of molecular biomarkers for the development of therapeutics tailored to an individual's disease (Collins, 2010). By moving beyond pathological phenotypes to better characterize disease and guide treatment, next-generation personalized medicine moves more deeply into the biology of the malady to refine diagnostics at a genotypic or molecular level. By understanding the unique biology of an individual's disease, in most cases a malignancy, personalized therapies can be directed against specific biological targets. This focus maximizes therapeutic effect, decreases the variance of therapeutic response through more accurate diagnostic classification, and minimizes side-effects.

In this chapter I suggest another domain where the metaphor of personalized medicine is apt and consider its application in the realm of neuropsychiatric disorders (Fins and Shapiro, 2014). To do this I invoke the example of emerging diagnostics, therapeutics and neuroethics informing the care of patients with disorders of consciousness (Giacino et al., 2014; Fins, 2019). These conditions comprise a range of brain states that span coma, the vegetative, and minimally conscious states. Each of these states have behavioural and biological characteristics that warrant review and consideration for our discussion of personalized medicine.

With disorders of consciousness, as with cancer biology, overt presentation and underlying biological mechanisms may be discordant. Patients

Joseph J. Fins, *Personalized medicine and disorders of consciousness* In: *Can precision medicine be personal; Can personalized medicine be precise?*. Edited by: Yechiel Michael Barilan, Margherita Brusa, and Aaron Ciechanover, Oxford University Press. © Oxford University Press 2022. DOI: 10.1093/oso/9780198863465.003.0010

presenting similarly at the bedside may have meaningful differences in their underlying neural circuitry. These distinctions, or the discordance between observed phenomenology and unobserved neurobiology, may have normative implications for the diagnosis and treatment of these conditions. Any genuine 'personalization' must grant a special status to conscious states and mental events.

To begin, let us review the prevailing nosology attendant to these brain states. When a patient loses consciousness, it can return without incident, as after fainting or anaesthesia. But in the cases of severe trauma, metabolic, or anoxic brain injury the duration and nature of impairment can linger. These conditions have historically been characterized diagnostically by their overt behavioural manifestations, that is responsiveness and neurological findings seen on bedside examination. To introduce a metaphor from genetics–so important to personalized medicine and cancer care therapeutics—what is seen at the bedside might be best thought of as a *neurological phenotype*.

This phenotype varies by brain state. For coma, it is characterized by an eyes-closed state of unresponsiveness. Coma, which is a transient and self-limited state of about 10–14 days, unless it is prolonged or induced by medication, either resolves or progresses to the vegetative state (Posner et al., 2007). The vegetative state, first described by Jennett and Plum (1972), is a state of wakeful unresponsiveness in which the eyes are open but there is no awareness of either self, other, or the environment. These patients have sleep–wake cycles and even a startle reflex but are unconscious. Depending upon the duration of the vegetative state it has been characterized as persistent or permanent. If it lasts for a month it is described as persistent vegetative state. If it lasts over three months following anoxic injury and twelve following traumatic injury it has been designated as permanent vegetative state (Multi-Society Task Force on PVS, 1994).

In 2018, these criteria were revised by the American Academy of Neurology, the American College of Rehabilitation Medicine as well as the National Institute on Disability, Independent Living, and Rehabilitation Research (Giacino et al., 2018). These guidelines notably altered the designation from the permanent vegetative state to the *chronic* vegetative state, since upwards of 20% of patients once thought to be permanently vegetative have the potential to migrate into the minimally conscious state (MCS) (Fins and Bernat, 2018).[1]

---

[1] This change in designation, as yet not fully appreciated by the bioethics or religious communities, could well have a bearing on how society reflects upon the right to die, in large part predicated upon the presumed futility and irreversibility of the vegetative state (Fins, 2006). While

MCS as a diagnostic condition was codified in 2002 as a state of consciousness in which patients show behavioural evidence of self, others or the environment (Giacino et al., 2002). They will say their name, reach for objects, respond when someone comes in the room but only do so episodically and intermittently. The variability of the MCS patient's phenotypic presentation of consciousness causes significant diagnostic challenges. When behavioural manifestations of consciousness are not demonstrated, the patient appears vegetative. A notable paper found that 41% of patients in chronic care following traumatic brain injury and thought to be in the vegetative state were actually in the minimally conscious state (Schnakers et al., 2009).

We now appreciate from neuroimaging data that the underlying biology of the vegetative and minimally conscious states are neurophysiologically distinct. Patients in the vegetative state have autonomic functions grounded in the brain stem without higher integrative function. If properly diagnosed, vegetative patients do not have intact distributed neural networks in their brains. When presented with stimuli, the usual processing networks that handle and interpret this input are not activated beyond the primary sensory areas (Laureys et al., 2002). Further integrative activity does not occur. This contrasts with the MCS patient where network responses, similar to unaffected individuals, can be sustained (Schiff et al., 2005). Notably, this allows MCS patients to experience pain or process language in a manner impossible for vegetative patients (Schankers et al., 2010). This calls for both universal pain precautions and a concerted effort to look for covert consciousness and provide assistive devices to aid in communication.

A failure to properly assess these patients may have profound ethical implications. If patients are diagnostically mischaracterized by phenotype as vegetative when they are in fact minimally conscious, they can either be thought to be insensate and in the vegetative state or non-communicative. Whereas these presumptions would adhere to vegetative patients, they would not to patients in MCS. Thus mischaracterized, MCS patients could be deprived pain medication and/or deprived of communication, compounding untreated pain with the tragedy of an isolation known only to them and not suspected by family or friends who believe them incapable of

---

this is true for the majority of the patients in the chronic vegetative state, the consequences of this change in designation from permanent to chronic could well have consequences in society outside of medicine (see also Fins and Bernat, 2018) since the right to die began in the vegetative state in cases like Quinlan and was later contested in cases involving other patients in the vegetative state (Cruzan and Schiavo; see Fins, 2006).

responsiveness (Fins, 2017). This has implications for what I have described as neuropalliative care (Fins and Pohl, 2015).

Herein lies the problem for a phenotypic characterization of patients with disorders of consciousness. It may be inaccurate and thus biologically misrepresentative. Whereas a vegetative and minimally conscious patient may have similar phenotypes, their *neurologic genotype*—grounded in the circuitry of their distinct underlying neural networks—may be quite distinct (Fins, 2007). Because MCS patients have intact neural networks, these patients are biologically different from vegetative patients with whom they can too easily, and sometimes ideologically (Fins and Plum, 2004), be conflated.

I described this disconnect between observed phenotype and neurologic genotype in patients with intact neural networks on neuroimaging and a vegetative examination (Owen et al., 2006), as evidence of a non-behavioural minimally conscious state (Fins and Schiff, 2010). If a patient activated appropriate regions in response to stimuli on neuroimaging, they could not logically fit the vegetative criteria of wakeful unresponsiveness. They *were* responsive, albeit at a level not observable by clinical examination. More recently, this phenomenon has been described as cognitive motor dissociation (CMD) (Schiff, 2015), where what is manifested motorically or behaviourally is not commensurate with the cognitive capabilities within the brain.

The challenge of the discordance represented by CMD is akin to that seen between phenotype and genotype in conventional discussions of personalized medicine. Those of us who have treated patients with cancer, *with seemingly the same disease* have had patients with variable outcomes. I make this analogy not to discount the epistemic and ontological differences between oncology and disorders of consciousness but rather because the origins of personalized medicine come out of oncology where patients with seemingly the same tumour have different outcomes despite similar phenotypes. The reason for this variability resides in genomic differences: differences that are accommodated in personalized medicine which treat a patient's genetic array and not just their phenotype.

This difference has made prognostication difficult. Why are these patients responding so differently to the same treatment? With the advent of genetic arrays, we now appreciate that cancers appearing to be the same diseases histologically were in fact biologically distinct. Considering cancer genetics, these conditions were distinct. They responded differently to treatment because they were different. With this realization, prognosis and treatment has evolved, with each informed by more sophisticated biomarkers.

In cancer, biology precision medicine is an emerging success story. As a resident in medicine at Memorial Sloan Kettering Cancer Center I recall treating patients with the dreaded acute pro-myelocytic leukaemia (APL) who did worse than their peers (Coombs et al., 2015). They had horrific bleeding diatheses and were paradoxically treated with heparin to quell this complication. Of all the leukaemia subtypes, this was the one to avoid. But today, with a deeper understanding of APL's biology and with the advent of targeted therapy with all-*trans* retinoic acid, this is one of the leukaemia sub-types with the best prognosis. There are, of course, other examples.

The challenge with disorders of consciousness is that we are several decades behind the progress made in cancer biology and personalized medicine. We are in a transition phase between behavioural criteria and a fuller understanding of the circuitry that might better characterize these conditions. With the advent of functional imaging, it almost is as if we had just discovered the microscope.

This presents a problem for prognostication and speaking with families as we make a transition between established behavioural metrics used at the bedside such as the Coma Recovery Scale-Revised (CRS-R) (Giacino et al., 2004) to neuroimaging criteria. For example, a patient may appear to be in the vegetative state but demonstrate volitional responses on functional neuroimaging. As yet, the criteria for diagnosis are based on behavioural criteria with only an emerging willingness to accept the role of neuroimaging in diagnostic refinement. This creates a challenge for diagnosis until the specificity and sensitivity of neuroimaging tests have more fully matured and are vetted. While a physiologic approach to understanding will be useful in the longer term, the CRS-R—the behavioural assessment tool—still has the best test characteristics of any diagnostic modality. The challenge is how to use this behavioural (phenotypic) tool in conjunction with new technologies like neuroimaging and EEG. In the long run, this combination will yield a powerful synergy, although it may produce confusion and mixed signals in the short term, making conversations with families especially challenging.

It is a colloquial saying to observe that knowledge is power. Our knowledge base has expanded enormously with the advent of neuroimaging and a way to peer into the brain, sometimes contradicting the ground 'truth' of bedside evaluation and what was taken to be reality, until we realized it was not as accurate as we once imagined. But the challenge is that we still know less than what we need to know, and the test characteristics of neuroimaging are not fully dispositive. Positive results can confirm covert consciousness, but their absence cannot refute it. This remains experimental and patients can present variably. So while we now know that covert consciousness is possible, this possibility prompts a whole new suite of questions with which we did not previously need to grapple. Here scientific advance presents us with a paradox: new knowledge may lead to increased uncertainty and be disempowering, since we now know more but are less certain of what we know when we have negative neuroimaging results and a good predicate for covert consciousness. Consider the example of covert consciousness. Before we had neuroimaging, we had no knowledge of its possibility. Our assessment was based on what we saw at the bedside and we believed what we saw. Now we are not so sure. Neuroimaging, as a technology, has expanded our knowledge and presented us with a normative problem for which we still do not have a reliable solution (Fins, 2011). As yet it provides answers and raises a host of new possibilities and responsibilities with respect to pain management and the provision of assistive devices to facilitate communication. An associated challenge posed by our advance in knowledge is presented by the newly designated chronic vegetative state. If 20% of patients once thought to be permanently vegetative have the prospect for additional recovery, how long is it prudent to wait? At what emotional costs for families and economic costs for society? We are in a period of tremendous intellectual advance but with this progress come new ethical challenges of which we were once perhaps blissfully unaware. Whereas previously clinicians thought they knew unconsciousness when they saw it, now they cannot be sure. Consciousness might lurk beneath the surface unrecognized. Or it could reappear with the tincture of time, and/ or novel therapeutics (Posner et al., 2019).

And this leads to personalized medicine and emerging therapeutics for this population. Like personalized medicine in cancer biology, which is grounded in generalizable pathways, neuromodulation is based on an emerging knowledge of the circuitry of consciousness and its disruptions. As certain molecular-based illnesses are caused by discrete pathologic abnormalities that operate within known pathways, neuropsychiatric conditions have dislocations within constituent circuits. Recently, the *mesocircuit* has been identified as a key circuit that plays a key role in consciousness and its disruption (Schiff, 2010).

Simply depicted, the mesocircuit links the thalamus with the brain stem below and the cortex above with the additional mediation of the basal ganglia. The brain stem provides arousal that is necessary but still insufficient for the achievement of consciousness. This requires the involvement of the integrative functions of the thalamus, which brings in the cerebral cortex. It is a widely projected modular system with cortico-thalamo-cortical pathways whose disruption provides physiologic opportunities for personalized therapeutic engagement.

In 2007, our group pioneered the manipulation of this system with the first use of deep brain stimulation (DBS) in the minimally conscious state (Schiff et al., 2007). Our work demonstrated the ability of intralaminar thalamic DBS to improve cognitive function in a severely brain-injured patient. In this proof of principle, first in humans, double-blind, cross-over study, a patient who had been in a stable MCS for years with an inability to communicate, eat by mouth and with impaired motor and postural control was able with deep brain stimulation to say six- or seven-word sentences and the first sixteen words of the American Pledge of Allegiance. He could go shopping with his mother and voice preferences for clothing (Fins, 2015). He could eat by mouth for the first time in six years, manage his secretions and had improved motor and postural tone.

From a normative point of view, his improved functionality led to what I have described as the restoration of agency *ex machina* (Schiff et al., 2009), the ability to participate in decisions about his care at the level of assent and dissent, demonstrating a degree of agency, an essential ingredient of agency. In my book (see Fins, 2015) I recount how in the aggregate this return of function led to the restoration of personhood, self-awareness and continuity in relationships consistent with one's 'former' self. On one occasion, the subject was crying. His mother assumed that he was depressed about his condition and asked him why he was crying. To her surprise, he replied, 'I am crying for Cory', his brother. As the narrative emerged, she told me who Cory was: 'That's his brother. That's his brother that does not come to see him, that's in denial'. And then she realized what this meant for her son's sense of being in the world and personhood. With a sense of amazement and surprise, she told me, 'And even I started crying because here he is, he's aware, he knows what's going on' (see Fins, 2015; Fins, 2020).

Like many people with disabilities, and most notably disorders of consciousness (Wright et al., 2018; Fins et al., 2020), the subject had been underestimated. Even his mother had to recalibrate her assumptions about his capabilities. He was aware of himself and his relationship to others. In the aggregate she told me that in her view his old self had returned, noting that, 'He's still Freedom', the nickname that he carried before his injury (see Fins, 2015). Certainly there is irony to the name in light of how he was liberated from the confinement of his disorder of consciousness with a neuroprosthetic device.

But the point is clear—with this technological assistance he regained person-hood and gave voice to agency.

Whereas conceptions of personhood vary from culture to culture—and this is a topic that has a complex history in philosophy—I would assert that critical dimensions of personhood, lost beneath the shadows of a disorder of consciousness, were restored through the emerging therapeutic capabilities of neuromodulation.

Like molecular-based medicine, the provision of emerging therapeutics—most still in the investigational stage—requires skilled diagnostic assess-ment that transcends bedside examination, and which is conversant with circuit-based derangements. Like molecular-based personalized medicine, we envision that a new nosology will emerge based on the therapeutic targets offered by distinct injuries. This would represent a transformational change in how these conditions are categorized, moving robustly from behavioural to circuit-based classification.

Until that occurs, and before we can assert generalizable patterns from a collected experience of engaging in this probative biology—itself a product of therapeutic investigations such as our work using deep brain stimulation in the injured brain (Fins, 2015)—we will be in a state of personalized medi-cine by default. We will try to apply incomplete data sets in a particularistic way to individual patients with unique injury patterns. This will lead to inter-ventions that appear as $N = 1$ cohorts (Schlaepfer and Fins, 2010).

This methodology, though essential at the frontiers of knowledge, makes the traditional metrics (and ethics) of randomized clinical trials impos-sible. It calls for a different regulatory approach and what the philosopher of science, Miriam Solomon, has described as a pluralistic way of knowing. She suggests that we need both methodological rigor as well knowledge of mechanism and the integration of personal narratives (Solomon, 2015; Fins, 2016). A key question is whether personalized medicine's integrative incorp-oration of plenty and diverse data embodies Solomon's vision of an integra-tive approach, or whether it is crude reductionism to 'big data' algorithms. The emerging nosology of disorders of consciousness will help elucidate how much attention should be given to phenomenology and clinical observation, at least in neuropsychiatric disorders.

There is the additional epistemic complication when considering neuro-psychiatric conditions. Whether one is attempting to restore or give voice to consciousness, as my colleagues and I have tried to do (Fins, 2016), to treat drug-resistant depression, as the neurologist Helen Mayberg has done also using deep brain stimulation (Mayberg et al., 2005), or to live with Parkinson disease, the question arises about the personal nature of the outcome (Dubiel, 2009; Crowell et al., 2015). What are the goals and who decides them is perhaps the ultimate question of personalized medicine and the continuity of the true self (Nyholm and O'Neill, 2016).

Fundamentally then, what we are witnessing in the demarcated clinical landscape of disorders of consciousness is an epistemic transition between what is observed and what is understood. This transition represents a move away from patterns of practice informed by decades, if not centuries, of diligent clinical observation, into a state of personalized care informed by the particularities of a patient's injuries. In time, these responses to 'personalized' therapeutics will be aggregated into generalizable findings applicable to groups of patients.

When this occurs, investigators will come up with a new nosology that replaces the descriptive ones upon which we have long relied. This new diagnostic classification will not solely be grounded in what is observed, but on the responsiveness of circuit derangements to neuromodulation. In this way we will improve diagnostics and prognostication, and refine therapeutics for this population.

If we achieve this objective we will have fulfilled an ancient mandate of medical inquiry as described by the great physician Sir William Osler, a founder of Johns Hopkins School of Medicine and late Regius Professor of Medicine at Oxford. In his classic "The Leaven of Science", Osler observed that, "The determination of structure with a view to the discovery of function has been the foundation of progress" (Osler, 1904).

I believe that this period of so-called personalized medicine is part of this Oslerian journey. It only seems personal because we need to make our discoveries one patient at a time. But this quest is in a long tradition of medical inquiry pushing back the shrouds of ignorance so that patients individually, and then collectively, will benefit from the foundation of progress that has sprung from our careful interrogation of function and its variance in disease and injury.

Dedication
The author dedicates this essay to the memory of John L. Battenfeld, M.D., whose love of the humanities informed the care of his patients.

11

Gender and personalized medicine

Methodological and ethical pitfalls

Marianne J. Legato

Introduction

The detailed description of the human DNA alphabet at the beginning of this century promised us astonishing new capabilities. Not the least of its consequences was a paradigm shift in the focus of healthcare. The medical community heralded genomic analysis as a novel and potent instrument which would allow us to assess an individual's susceptibility to illness, to develop tailored methods for that person to prevent disease, to monitor the patient for the first evidence of illness if it appeared and to fashion specific, effective treatment once it occurred. This concept of personalized/precision medicine was born out of the immediate explosive enthusiasm for the nascent biology of the genome. Altman (2015) commented in his foreword, 'Biology's Love Affair with the Genome':

> Postgenomics is the unavoidable consequence of an intense love affair between biomedical scientists and the human genome. The breakneck speed of the courtship slows to a more reasoned set of discussion, negotiations, and settings of expectation. Some love affairs do not survive these adjustments, but others transition to a lifelong shared adventure.

The initial euphoria has been replaced by the realization that the task of delineating the molecular machinery that establishes and maintains the phenotype is almost intractably complicated. In their landmark report describing the sequence of the human genome, Venter and his colleagues warned that this achievement was only 'the first level of understanding of the genome' (Venter, 2001, p. 1348). One of the first issues to be addressed was the totally unexpected finding of the relatively small number of genes

Marianne J. Legato, *Gender and personalized medicine* In: *Can precision medicine be personal; Can personalized medicine be precise?*. Edited by: Yechiel Michael Barilan, Margherita Brusa, and Aaron Ciechanover, Oxford University Press. © Oxford University Press 2022. DOI: 10.1093/oso/9780198863465.003.0011

in the human: 26,000–38,000, compared with earlier estimates of 50,000 to >140,000. Moreover, any two individuals are more than 99.9% identical in genomic sequence: 'the glorious differences among individuals in our species that can be attributed to genes falls in a mere 0.1% of the sequence'.

The authors warned that the myriad features and complexity of human physiology could not be explained by the seductive but simplistic concept of gene function. Instead, they offered the concept of a complex system of genetic regulation:

Neither gene number, neuron number nor number of cell types correlates in any meaningful manner with even simplistic measures of structural or behavioral complexity. Nor would they be expected to: this is the realm of nonlinearities and epigenesis.

Finally, they warned:

There are two fallacies to be avoided: determinism, the idea that all characteristics of the person are hard-wired by the genome; and reductionism, the view that with complete knowledge of the human genome sequence, it is only a matter of time before our understanding of a gene's function and interactions will provide a complete causal description of human variability.

In summary, gene regulation is the basis of phenotypic variation and adaptation to the environment.

Limitations and caveats of personalized medicine

Personalized medicine is a laudable goal, but in fact attempts at an accurate assessment of each unique individual will inevitably be imperfect and incomplete. They will be shaped by the sophistication and world view of the observer/scientist, the effectiveness of the technologies employed to collect and integrate the data, and the impact of environment, biological sex, and culture on gene expression in each person, some of which will be imperfectly understood or not appreciated at all.

We are at the point in genomic science where we have acquired huge databases of genomic information about individuals; Siminoff (2017) reported that reasonably current sources document more than 500 million biospecimens containing genetic material stored in public and private databanks. But, as Venter (2010, p. 676) comments:

> The generation of genomic data will have little value without corresponding phenotypic information about individuals' observable characteristics and computational tools for linking the two.

He correctly points out that collecting, much less keeping confidential, adequately detailed data about each participant in the genome databases will be as complicated and challenging (not to mention perhaps much more complex) than defining the molecular structure of the genome has been.

Several formidable obstacles exist to making the data relevant to patient care:

- Although the genome is identical in 99.4% of all humans above (Siminoff 2017), the variability among them is due in great measure to the complex process of *genomic regulation*. This aspect of systems biology is only beginning to be explored and is a subject of intense interest in the biomedical community.

- Linking genomic to phenotypic data is an enormous task: how the environment, biological sex, hormones, and age affect the genetic phenomena underlying function is only beginning to be considered. The computational tools necessary to link genomic information comprehensively and accurately to the phenotype have yet to be developed.

- In any consideration of how biological sex affects gene expression, the subsets of molecular biology that characterize *transitional forms* between male and female must be taken into consideration, adding another significant element of complexity to interpreting genomic data.

- Working with lay and scientific members of *under-represented groups* is essential not only to make the science more accurate but to provide access to its benefits to underserved populations of the world.

- Issues of *confidentiality* and the implications of sharing genetic information not only with the individual person but with family members, environmental agencies such as insurance companies and political systems, will require new regulatory systems that are universally acceptable to those who utilize the data: parameters for guidance and regulation have yet to be defined.

Does sex matter in biomedical investigation?

Women as subjects of clinical investigation

The emerging participation of, and attention to, women rather than an exclusive concentration on male subjects in clinical investigation is one of the hardest fought and interesting phenomena of the last century.

Fuelled by the two world wars of the 20th century, the societal view of women as less equipped physically and mentally than men to deal with the challenges and realities of the world outside the home radically changed. The 20th century American woman insisted on and achieved the right to vote, assumed leadership roles in industry and healthcare institutions, and filled jobs from farms and factories to governance on every level of the economy. Women emerged mid-century with a new awareness and affirmation of their expanded competence and on every front began to advance the cause of a corrected societal view of their newly acknowledged abilities. *Pari passu*, they questioned the view of themselves as physiologically identical with men except for reproductive biology, refuting the bikini view of women's health, where breast and pelvic biology were the only elements of gender-specific care.

By 1970, American feminism was pushing forward the principle that society needs a complete revision of its views on women and gender. This extended to the growing awareness that female physiology was distinctly different from that of men in ways that were just beginning to become apparent. The Boston Women's Health Book Collective published the landmark *Our Bodies, Ourselves* (1973), and petitioned the United States Public Health Service to survey what was actually known about female physiology. The Public Health Service Task Force on Women's Health Issues responded with the rather astonishing observation that, apart from reproductive biology, almost no biomedical investigations had included women (Kirschstein and Merritt, 1985). The second wave of feminism in 1970 fuelled an extensive concentration on the unique aspects of women's biology. This prompted a response in July 1989 from the National Institutes of Health (NIH) to encourage the inclusion of women in clinical research, but, by 1990, the General Accounting Office reported that the policy had been poorly applied ('History of Women's Participation in Clinical Research'). Three years later, the American Congress passed the NIH Revitalization Act (Public Law, 103-43) requiring the inclusion of women and minorities in medical research. Notably, it has been politics energized by grassroot activism rather than by scientific methodology that determined this inclusion. However, it did not include any recommendation about including females in animal research; that was to come much later, from the National Institutes of Health itself (Clayton and Collins, 2014).

It is important to point out that the exclusion of women before this new era, which had its beginnings in the last quarter of the 20th century, was not the result of regarding women as less valuable than men. In fact, it was essentially an economic and pragmatic decision: men, largely recruited at the end of World War II from Veterans' Administration hospitals, were (correctly) assumed to be likely to accept the risks of being part of clinical studies, and furthermore were regarded as physiologically more stable than women because of their freedom from cyclic hormonal fluctuations. More profoundly, perhaps, the notion of women's primary value being in their ability to conceive and bear children was deeply rooted in society's view of why they should be protected from the risks of biomedical investigation. Paradoxically, the consequences for male reproductive biology were not really counted an important factor, reflecting the view of men as primarily important because of their ubiquitous role in the construction, governance, and maintenance of the societies in which they lived. Conversely, and equally important, however, was the injustice of having men exclusively bear the risk of participating in biomedical investigation while women were exempted from that risk. The Belmont Report (National Commission for the Protection of Human Subjects of Biomedical and Behavioural Research, 1979, p. 5) highlighted the importance of justice in the participation of individuals in research:

> Who ought to receive the benefits of research and bear its burdens? This is a question of justice, in the sense of 'fairness in distribution' or 'what is deserved'. An injustice occurs . . . when some burden is imposed unduly.

It is fascinating that this relatively recent summary of the preferred ethical and moral laws that govern research on humans includes an exhaustive overview of what should be considered in selection of subjects, but *never comments upon or urges the inclusion of females* in clinical trials. The medical community and society at large and at every level had made and was continuing to make the unjustified assumption that male and female physiology were essentially identical and that data from men could be applied to women without direct testing/verification. Similarly, the Declaration of Helsinki, first conceptualized in 1964, was revised eight times during the next 38 years, the last version issued in 2013 (World Medical Association, 2013). There is no mention whatever about the importance of including both sexes wherever possible in clinical investigation, and the reference to animals is a single sentence: 'The welfare of animals used for research must be respected.' This declaration is formulated in an abstracted agender jargon, referring to 'human subjects'. No reference is made to children or to any other group of people.

Finally, in 2016, the Council for International Organizations of Medical Sciences (CIOMS) published Guideline 18: 'Women as Research Participants', and carefully defined the importance of women's inclusion in health-related research. Appropriately, they cautioned that women of child-bearing potential deserved specific information about potential risk to the foetus if they became pregnant during a research trial.

Gender-specific clinical investigation has produced a completely unexpected cornucopia of rich detail about the striking differences in how disease affects the sexes. While the interest in gender-specific medicine began in the sex-specific aspects of cardiovascular disease, it soon became obvious that other aspects of human physiology and the pathology of disease might also be sex-specific. The expanding new science of gender-specific medicine began to take shape (Legato, 2016).

The prestigious voice of the National Academy of Sciences formally endorsed the concept of gender-specific medicine with the publication of its monograph in 'Exploring the Biological Contributions to Human Health. Does Sex Matter?' The conclusion of the committee was incontrovertible (Wizemann and Pardue, 2001):

> Being male or female is an important fundamental variable that should be considered when designing and analyzing basic and clinical research . . .

Probably the most important contribution of the monograph to biomedical research guidelines was the principle that all scientific reports should determine and disclose the sex of origin of biological research materials—not only that of human subjects. Like all new ideas, the directive was only slowly accepted and implemented; research protocols on animals and even simpler preparations like cell lines were not constructed to assess the impact of sex on the data. Writing in 2012, Virginia Miller commented that less than 40% of studies using animals identified the sex of the experimental material in spite of repeated urgings of NIH leadership to acknowledge the importance of including sex as a significant variable in experimental protocols at all levels of investigation (Miller, 2012).

The importance of sex in genomic research

Since the discovery of the chromosomes, the prevailing scientific theory of sex is rooted in the genes. However, the search for the multiple factors that operate in dynamic synchrony to establish and maintain sexual identity and behaviour continues to evolve. In fact, sexually dimorphic traits are the consequence of sex-specific genomic expression rather than from a difference of gene structure itself in males and females. The identification of a sex-biased gene is achieved by transcriptome profiling using microarrays constructed with the use of commercially available genomic sequence data or from reads of expressed sequence tags (ETA) using DNA libraries made specifically from males or females (Elligren and Parsch, 2007). It is now widely appreciated that sex-biased expression of genes is ubiquitous.

Echoing the conviction of scholars who work in the field of gender-specific medicine, Arthur Arnold appropriately points out that in addition to its impact on normal physiology, biological sex may actually promote a dimorphic response to illness, resulting in protection from, or increased vulnerability to, any given disease (Arnold and Lusis, 2012). This is more than evident clinically; we have known for centuries that the incidence of specific diseases is often strikingly unequal in the two sexes. Furthermore, the course and manifestations of the same illness may vary significantly between the sexes.

Arnold reasons that identifying those gender-specific elements that so influence the pathophysiology of disease might present opportunities for novel interventions to prevent and/or treat patients. Appropriately, he urges not only the focused study of female physiology—so significantly different than that of males—but also simultaneous comparison of data from both sexes.

What factors at the genomic level determine sex?

What progress have we made in the postgenomic era on identifying the mechanisms of how sex is established and maintained in humans? Sarah Richardson is a prime mover in the search for the elements that collaborate to establish and maintain sexual identity. She points out repeatedly that our historical concentration on the X and Y chromosomes as 'the two pillars of sexual identity' is not only simplistic but, in fact, wrong. Richardson (2012) writes:

. . . scientists must work to develop alternative models of the relationship between the X and sex . . . the X has been overburdened with explaining female biology and sex differences. (The X) has become a highly gendered screen upon which cultural theories of sex and gender difference have been projected throughout the twentieth century and up to the present day.

There remains a persistent assumption that the sex chromosomes are, in fact, the locus of genes responsible for sex differences. Geneticists, clinicians, and science reporters commonly assert that genes on the sex chromosomes will elucidate the genetic substrate of sexual dimorphism. According to this view, genes or gene processes involved in primary and secondary sex characteristics should be located, or concentrated, on the sex chromosomes. This approach reflects the expectation and key assumption that the genotypic dimorphism of the sex chromosomes is the biological substrate of, or controls, the phenotypic dimorphism between male- and female-bodied individuals that we observe. This assumption, however, is not valid.

In her discussion about how wrong it is to label the X the 'female chromosome', she reminds us that the X, unlike the Y chromosome, with its singular ability to flip the genetic switch for male development, has no similar capacity to initiate female sex development. The X, moreover, has a large collection of male sperm genes (Richardson, 2012).

Richardson pursues what she calls 'the search for male and female in the human genome', summarizing the data that challenge the exclusive role of the X and Y chromosomes in creating male and female and emphasizing the role of the myriad other elements in determining sex. On the other hand, Richardson (2012) seems to infer that both chromosomes exert a 'critical switch' that begins the trajectory to either masculinity or femininity. In the case of the X chromosome this may involve the triggering of other crucially important autosomal genes:

Sex chromosomes evolved as a sex-determining mechanism: they carry a critical switch in a larger pathway of genes that determine sex. Genes with sex-specific fitness effects need not be located on the sex chromosomes. There are many examples of autosomal genes critical to sex determination and sex differentiation.

Not only does Richardson recast the roles of the X and Y chromosome, but she also cites the importance of Arnold and McCarthy's work (McCarthy and Arnold, 2011) in emphasizing the impact of sex chromosomal genes on physiology, independently of hormones. She sums up the current view of what shapes sex:

> Whereas the genetics of sex was once largely limited to the sex chromosomes and gonad-determining genes, this new account of sex envisions the whole genome is imbued with sex-specific processes and sex differences in gene expression . . . in this new account, genes emerge as the coequal and even primary actors in sex differences, not second fiddle to the sex hormones. Richardson SS (2013, p. 214)

As McCarthy et al. (2009) put it, 'despite the supremacy of gonadal steroid hormones, the battle of the sexes is ultimately fought in the DNA'. Ellegren and Parsch (2007) reinforce this principle and remind us that the male and female genomes differ by only a few genes. They reiterate the fact that most sexually dimorphic traits result from sex-specific expression of genes that are present in both sexes.

Arnold et al. (2012), like Richardson, support a primary role of the X and Y chromosomes in establishing the biological sex of the individual:

> The primary sex-determining factors are encoded by the sex chromosomes, because all sex differences start with the sex chromosomes at some point in life. (They) are the only factors that differ in the male and female zygote, and thus they are the factors that give rise to all downstream sex differences thereafter.

That having been said, Arnold et al. (2012) articulate a much more expanded view of the maintenance of human sexual identity and behaviour, describing 'complex intersecting causal pathways in which genes and their products form networks of interactions as they control and are controlled by each other'; the genes 'pulsate with activity, stimulating and inhibiting each other, creating a dynamic net of interactions'. Arnold and Lusis attribute sex differences to the action of these gene networks and they name the entire, complex and interacting system the 'sexome' (Arnold and Lusis, 2012). In addition, they advocate a focus on identifying these sex-biasing factors to foster more accurate prevention of disease and the treatment of human pathophysiology.

Investigators universally accept the fact that sexual dimorphism is significantly fashioned by gonadal hormones but all too often consider them the only force delineating the unique characteristics of the two sexes. In fact, the action of the sex chromosomes themselves is a separate and essential agent in shaping the phenotype. Furthermore, as Ober et al. (2008) point out—agreeing with Richardson—autosomal genes (heretofore assumed to be the same in both sexes) are often the agent of differences between males and females (p. 913):

> Although genes with sex-biased expression are enriched on the sex chromosomes, thousands of sex-biased genes are also found on the autosomes . . . . Although the DNA sequence, gene structure and frequency of polymorphisms on the autosome do not differ between males and females, the regulatory genome is sexually dimorphic.

Yang et al.'s (2006) observations confirmed this over a decade ago. Her group reported that indeed, even at the molecular level, sex-specific genes are expressed in the body's tissues with small but significant differences in the proteins they produce. They comment (pp. 995–1000):

> Here, we report surprisingly wide-spread sexually dimorphic gene expression . . . . Thousands of genes showed dimorphism in muscle, fat and liver; hundreds in the brain.

This striking abundance of sex-specific gene expression explains the multitude of differences in phenotype between male and female *in spite of nearly identical genome sequences.* Importantly, the sexually specific genes were highly tissue specific. Moreover, in addition to the X and Y chromosomes, Yang's group also affirmed the existence of tissue-specific, sex-specific autosomes.

Usefulness of genomic research for the public health

The enthusiasm for genomic investigation to produce significant health benefits for the human population is not shared by the whole scientific community. Joyner and Paneth (2019) point out that Francis Collins' prediction that personalized medicine would have a transformative impact on the treatment of disease has simply not been realized. They offer six challenges to the enormous resources being devoted to genomic research (Joyner and Paneth, 2015), citing:

- A failure of personalized medicine to contribute significantly to predict vulnerability to disease.
- Even with two successes in disease gene identification (BRCA1/2 and CF), there has not been a significant improvement in the treatment of breast cancer, and that offered to cystic fibrosis patients is useful only in patients with specific CFTR mutations.
- With our new awareness of the impact of the environment on gene expression, the harvesting of accurate and complete data about an individual's social/cultural context has not been implemented in a consideration of therapeutic or preventive choices for the patient.
- There is a danger of conflict of interest at the institutional level in the efforts to obtain federal, venture capital or individual philanthropic funding based on predictions about the value of personalized medicine as opposed to the benefit of traditional epidemiologic projects such as the Framingham Study which 'have specific hypotheses that inform data collection and carefully assess prespecified exposures and outcomes in a standardized fashion'.
- The expense of personalized medicine will be much greater than the interventions to prevent and treat disease that are applied broadly to populations.
- Historically, improvements in health have come from general improvement in socio-economic conditions or from programmes targeted broadly to entire populations such as improved sanitation, mass immunization, and tobacco control.

In spite of these caveats about the value of the financial and temporal support we are using to support genomic research in an effort to improve health, we should continue to pursue a more accurate understanding of the difference in vulnerability to, and in the experience of, disease as a consequence of biological sex. The history of gender-specific medicine documents the value of more soundly focused prevention and treatment recommendations for men and women.

The harvest of genomic investigation is not always insignificant: there are, in fact, some striking success stories. Its application to cancer has been one of the most valuable: the data reinforce the essential importance of biological sex in the clinical experience of malignancy. Whereas most studies focus on protein-coding regions of the genome, Li et al. (2019) investigated autosomal non-coding regions and reported sex-based mutations in the first pan-cancer analysis of sex differences in whole genomes of 1,983 malignant tumours, demonstrating functional consequences of genomic sex biases on the transcriptome and tumorigenesis.

The sexual dimorphism in gene expression contributes to, and helps explain, sex-specific susceptibility to disease. That susceptibility, moreover, is often tightly related to developmental stage and thus by inference to hormonal profiles that change significantly over the lifetime. In a recent *bioRχiv* preprint, Natri et al. (2018) lament the omission of obvious regulatory differences between men and women in the risk for, and features of, the same cancers. In spite of the overwhelming data that the susceptibility and clinical features of the same cancers are sex-specific, they comment on the lack of a consideration of sex differences in the development of cancer therapies.

In a landmark assessment of the impact of biological sex on susceptibility to disease, Westergaard et al. (2019) published the data from the entire hospitalized population of Denmark (almost seven million hospital admissions) over a 21-year period; 48.2% of these were women. Although both sexes shared many of the same diseases, the incidence, age of onset, longitudinal pattern, and frequency of associated illness were distinctly different between male and female patients. The authors emphasized that their study was essentially a documentation of sex differences *irrespective of mechanistic molecular causes, links to differences in environmental exposures, or biases in the healthcare system.* The comprehensive and detailed nature of the study makes it a uniquely valuable resource to correlate genomic data with sex-specific clinical medicine. In a recent communication, Dr. Soren Brunak, research director of the Novo Nordisk Foundation Center for Protein Research, reported that Denmark has just passed a new law that 'established a State-run National Genome Center . . . that will host, generate and oversee genome data in a country wide fashion' (Petrone, 2018). He will head the Research and Infrastructure Committee of the Center and writes: 'These data will be linkable to all the data in the *Nature Comm.* Paper, following application approval' (personal communication, 2019). Such innovative and targeted correlations are important: they relate genomic information to clinical data and in this case can significantly enrich our appreciation of the impact of sex on genomic function.

Gilks et al. (2014) have constructed a summary table of single nucleotide polymorphisms with sex-dependent effects on human phenotypes identified through hereditability studies and genome-wide association studies. Stulp et al. (2012) explore what they term intra-locus sexual conflict that occurs across a single locus, with different consequences for each sex. They cite the example of body height, i.e. in general, men are taller than women despite height being controlled by the same molecular genetic configuration.

Not only is our conceptual view of sex differences changing, so is our ability to generate the science to support those perspectives. The eight-year-old Common Fund's Genotype-Tissue Expression (GTEx) Program is expanding our awareness of the significance—and mechanisms—regulating gene expression. It is dedicated to studying all the factors that govern gene expression and is an important step to integrating the sex in the regulation of gene activity. The GTEx consortium operates on the principle that although genome-wide association studies have identified thousands of loci for common diseases, the mechanisms of precisely how those loci are related to disease remain unknown. A much more fruitful approach to solving the issue is *to concentrate on the impact of regulatory regions in the genome*, since most associated variants are not correlated with protein-coded changes. Essentially, the programme proposes a bi-obank containing samples of the major tissues of 1000 deceased adults. The complexities of recruitment and appropriate informed consent of donors and their families are challenging. (For example, the responsibility of investigators who utilize donor tissue to return information that might not only interest, but affect the health of donor relatives, must be defined. Most importantly, how well and for how long after death post-mortem data would correlate with the physiology of living tissue still needs assessment.)

One of the most elegant examples of how the GTEx can yield sex-specific information with an unprecedented degree of sophistication is the recent paper from Tukiainen et al. (2017). The investigators established that incomplete inactivation of the X chromosome in female cells varies between individual cells and extends to population levels across a wide range of tissues. They reported that at least 60 genes were sex-biased due to escape from XCI and they hypothesized that 'escape' elements are contributing to sex-specific differences in health and disease.

Collaboration between the Office of Research on Women's Health and the Common Fund was highlighted in a joint 2017 workshop. The large databases built by the Common Fund contain a wide assortment of biomedical information on both sexes that is being mined and supported by supplemental grants to support research that considers sex an important variable in human physiology.

The sex-specific relationship between gene expression and experience

One of the most powerful components of epigenetic science is the proposal that epigenetic mechanisms translate experiential and environmental factors to the genome, modulating the phenotype in a sex-specific manner. Some of these changes are permanent and can even be inherited by at least two successive generations (Liu et al., 2008; Szyf et al., 2008; Morgan and Bale, 2011; Nilsson et al., 2012; MacLeod, 2016; Wang et al., 2017).

This is one of the most powerful concepts in biology: over the last decade, the conceptual division between the terms (biological) sex and gender has largely disappeared. Since the 1950s, 'sex' was considered the hard-wired biological characteristic of the organism, while 'gender' encompassed all the environmental experiences/influences that shaped human behaviour, which is in many respects fluid and profoundly impacted throughout the life span.

Some aspects of different cultural values and perspectives on the human condition (particularly with regard to the respective societal roles of *and the widely different value placed on* men and women) are virtually impossible to quantify but profoundly impact health and treatment modalities (Moazam, 2019). For example, Jafarey and Moazam (2010, p. 358) report one trenchant comment from a Pakistani student:

> Gender 'images' are created by society. . . and 'we spend our lives living up to those images.

The first years of seminars at the Office of Research on Women's Health at the National Institutes of Health were marred with striking confrontations between molecular biologists and experts in the social and anthropologic disciplines; the latter repeatedly criticized the biologist's disregard for the role of environment, culture, and experience in fashioning the human phenotype. It is now evident that the environmental sea in which the individual lives out his or her life modifies molecular biology by means of epigenetic modifications that control gene expression.

In a very real sense, then, genomic behaviour is plastic, changing throughout the life span in response to the individual's age and sex-specific physiology as well as his/her unique environmental experiences. Some of these changes are not only permanent but are passed on to at least two subsequent generations, often in a sex-specific manner. Thus, this new, expanded concept of personalized medicine provides us with a final common path for what were considered two quite separate modalities. It not only presents a compelling explanation of how we are able to adapt and compensate for what happens to us in the exterior world, but it makes apparent that sex-specific prevention and treatment of disease is urgently needed. It also explains the mechanisms by which we are the heirs of the experiences of previous generations.

Scott Gilbert (2018 p 42), an expert and historian of developmental biology, offers a new construct called ecological developmental biology:

> This is a science where organisms and environments possess agency, and the genome is both passive and active . . . . This is not an approach against genetics or big data. Rather, it demands a broadening of the scientific portfolio, so that other perspectives are also included in biological funding and as appropriate biological explanations. It argues that . . . the physical organism is critically important and that the environmental context plays a large role in gene expression.

Springer et al. (2012, p. 1818) echo this view:

> We are not arguing against any biologically based male–female differences. Rather we are arguing that the vast majority of male–female health differences are due to the effects of the irreducibly entangled phenomenon of 'sex/gender' and therefore this entanglement should be theorized, modeled and assumed until proven otherwise.

Biology is profoundly and dynamically impacted by the sea of environmental experiences in which the individual is immersed. The epigenetic responses to those experiences are sex-specific. Furthermore, if the individual is exposed to environmental phenomena at a critical time, the epigenetic modifications may not be temporary and reversible but fixed and inherited by subsequent generations. Thus, our focus turns not to gene *structure* as the determinant of sexual identity and behaviour, but to how the same genes are *regulated in a sex-specific* way by epigenesis: chemical modifications that impact the gene product without altering the structure of DNA itself. Thus, the leading science of the present search for sexual identity is that of epigenetics.

The blending of biologically determined sex and the individual response to experience from intrauterine development to old age through epigenetic modifications of gene expression is one of the most important discoveries of the postgenomic era. It adds layers of complexity to the study of the factors in the genome that are impacted by the environment. Hence, personalized medicine requires a meticulous documentation of physical, social, cultural and environmental factors that shape the individual person. Correlating the genomic sequence with physiology and pathophysiology without that documentation will be imperfect.

Niedzwiecki et al. have made an important proposal for a comprehensive approach to measuring the complex environmental factors that impact health. They expand Wild's concept of the *exposome*, 'a paradigm involving the study of the health effects of cumulative environmental exposures and concomitant biological responses from conception until death' (Niedzwiecki et al., 2018, p. 108).

They define the exposome as 'the entire set of environmental exposures throughout the life course', thus including exposures of all types. These are monitored outside the body or by biosamples from the individual. Essentially, the exposome is, as they term it, 'an environmental complement to the genome.' Miller, one of the most prominent authorities on the exposome, summarizes the biological consequences of the challenge of the environment. They include metabolic changes, protein modifications, DNA mutations and adducts, epigenetic alterations and perturbations of the microbiome (Miller and Jones, 2014).

The marriage of individualized, personalized genomic analysis with a global assessment of the thousands of chemicals to which a person has been exposed will require what the authors call 'a grand challenge for analytical chemistry, clinical science, precision medicine, epidemiology, toxicology and exposure science' (Niedzwiecki et al., 2005, p. 119). They cite high-resolution metabolomics (HRM) as the most promising analytical technology for an exosome platform underlying the observations of individual genomic analysis. Essentially, this concept, which turns from the molecular biology of the genome to assessing the myriad of environmental features to which specific populations are exposed, addresses the concerns of Joyner and Paneth about our investment in genomic science. It is an approach which unites 'precision medicine to population-based public health'.

Does the impact of the environment/socio-economic circumstances/culture erase differences in male and female behaviour and competence?

Several authorities have suggested that the submersion of the individual in the environment can mute differences between the sexes, particularly in the brain. Joel advances the notion that sex differences in the adult brain might be the result of sex differences in life experience (Joel, 2011; Joel and Fausto-Sterling, 2015). She observes that in most documented sex differences in the brain there is considerable overlap, most marked in the areas associated with behaviour, emotion, and cognition. Reports of sexual dimorphism in brain structure consistently document significant overlap between men and women (Garcia-Falgueras et al., 2011).

The notion that epigenetic modifications produced by environmental experience may produce a biological plasticity is a compelling way to marry the sexed molecular biology of the individual with the impact of the changing environment, both internal and external. Indeed, since the explosion of research in epigenetic science, it is clear that environmental experience has an impact on genomic function which can endure throughout the lifetime of the organism and even be transmitted to at least one subsequent generation. Richardson (2017) recognizes the power of the suggestion, but reminds us that epigenetic plasticity itself is sexually dimorphic and that it modulates a sex-specific response to the environment. Thus the question of how sex (a clearly complex situation dependent not only on the X and Y chromosomes, but on a whole variety of factors both internal and external) impacts function in any single individual is probably one of the most difficult to answer.

Transitional forms/variations in human sexual identity

The realization that sexual identity exists on a continuum rather than be relegated to a simple binary classification renders the assessment of sex as a factor in interpreting the genome at the personal level quite challenging.

Indeed, it is essential to explore the molecular biology of sex determination in a new context that offers a basis for the variation in sexual expression that is so evident in the human condition. Fausto-Sterling (1989, p. 330) makes a trenchant statement about the disconnect between the biological view of sex as a rigid dyadic classification and the reality of sexual variations in the population:

> . . . biologists' inability to break away from a strict binary account of male and female has led them to ignore data which are better accounted for in approaches which accept the existence of intermediate states of sexuality.

Richardson (2013) echoes the difficulty in reconciling the binary view she calls the 'twoness of sex', i.e. the fact that two different sexes are required for reproduction with the fact of a multiplicity of transitional forms of sexual identity and behaviour in society. She introduces the notion of sex as a 'dynamic dyad' toward what she calls a gene's eye view of collaboration, interaction, exchange, and interdependence between the sexes:

> In these ways, the concept of sex as a dynamic dyadic kind advances a picture of sexual diversity quite different from the type of reductionist and ideological binary thinking about sex that often characterizes both popular and scientific discourse. A critical conceptual shift in thinking of this sort about the ontology of genomic sex will be required as we approach the genomic and postgenomic age of sexual science. (p. 199)

Vilain emphasizes the usefulness of this concept: he points out that sex determination is not the result of one isolated element, like the SRY gene, which, although it is a pro male factor, is not the only entity involved in maleness: its action may be opposed, for example, by WNT4 and DAX1 (located on the X chromosome) genes. Vilain (2003) comments that sex determination consists in the interaction of a complex network of genes, many of them still unknown. It is their relative doses that shape the sex of the individual.

Clearly there is a sequence of variations in sexual identity that occurs at the genomic level and which, moreover, can be significantly altered by the epigenetic impact of hormonal and environmental factors (Legato, 2020). Nevertheless, from the point of view of the clinician, a macro view of humanity separated into two—albeit heterogenous—classes, i.e. male and female, has been a useful heuristic. As our insights progress into genomic medicine and the complex nature of how sex is determined and how sex modifies gene expression, we will enter a more precise and useful era in patient care. One of the most important aspects of such progress is that it will resolve the mechanisms of how transitional forms of sexuality are formed and place them appropriately on the sequence of variations in sexual identity.

Conclusion

The description and exploration of the genome is arguably one of humanity's most important accomplishments. In an ever-widening perception of how the collection of genes assembled on the double helix of DNA is regulated, we have incontrovertible proof of the central and ubiquitous importance of gender in understanding many aspects of human biology and pathology.

Moreover, the separation of biological sex and the effect of the environment on the living organism is a defunct concept. The epigenetic modification of the gene by the environment makes gender a complex factor that integrates both entities in a final common pathway. To exclude a consideration of gender in planning biomedical research at all levels, from the whole animal to the molecular biology of the components of single cells, is an indefensible intellectual error.

12

Potential challenges to doctor–patient trust posed by personalized medicine

Shlomo Cohen

Introduction

Personalized medicine is widely recognized as an important vision for medicine. Yet despite the impressive increase in the number of clinically useful molecular diagnostics and targeted therapies, and despite significant progress in precision treatment in the fields of rare diseases and oncology, the routine incorporation of personalized medicine into clinical practice is still mostly a future promise—indeed, personalized medicine is usually not even discussed at the point of care (Evans et al., 2011; Miller et al., 2016; Pathak, 2018). The still very modest incorporation of personalized medicine into the clinic is mirrored in ethical reflection on the subject, which has thus far given little attention to the clinical ethics aspects of doctor–patient interaction under personalized medicine. This chapter examines the possible risks that personalized medicine poses to trust in the relations between patients and doctors.

Trust and the doctor–patient relationship

Trust refers to our faith in others that they can be relied upon, that they will not betray our expectations (Baier, 1986). As one of the most basic interpersonal attitudes, it arguably defies definition in the proper sense—indeed, we may recursively, yet helpfully, explicate 'trust' as our faith that others will prove trustworthy.

Trust is of fundamental importance to human relationships and to social functioning. Were trust to be absent, people would not be able to interact with others even at a superficial level or carry out cooperative activities;

Shlomo Cohen, *Potential challenges to doctor–patient trust posed by personalized medicine* In: *Can precision medicine be personal; Can personalized medicine be precise?*. Edited by: Yechiel Michael Barilan, Margherita Brusa, and Aaron Ciechanover, Oxford University Press. © Oxford University Press 2022.
DOI: 10.1093/oso/9780198863465.003.0012

societies would then disintegrate into a Hobbesian 'state of nature'. Trust encourages us to see and respect one another as moral agents. In a climate of trust, intimate and professional relationships can take shape. Trust allows us to form relationships with people and to depend on them—for love, for advice, or for help; it thus redeems us from the poverty of exclusive self-reliance as well as from existential solitude.

Trust is of great importance to the doctor–patient relationship. 'Sick people have always had a particular need for trust, because to fall ill implies a loss of trust in yourself, in your body, in your social role, in your future. This loss of trust fortifies the need to trust others; among them, the doctor.' (Fugelli, 2001, p. 576). Both due to their vulnerability and to the asymmetrical knowledge of medicine, patients tend to trust doctors, transferring to them the power to act in the patients' best interest (Calnan and Rowe, 2008). Numerous studies have shown that patient trust has an impact on patient satisfaction, adherence to treatment, continuation of doctor–patient relationship, improvement in self-reported health, and other parameters of medical care (Hall et al., 2001; Skirbekk et al., 2011). Thus, acting in ways that justify and perpetuate patient trust is not only ethically right but also of great importance for the quality and success of healthcare.

Trust has two main forms: interpersonal trust and social trust. In the first, people place their trust in other individuals, such as in their physician; in the second, the objects of trust are collective institutions, such as a healthcare system. Social trust in the healthcare system and interpersonal trust in physicians diverge in important ways but they are also mutually supportive. Patients who trust their physicians will generally tend to worry less about the trustworthiness of the system; similarly, trust in one's physicians can flow from confidence in the competence and commitment of the institutions with which they are affiliated (Mechanic and Schlesinger, 1996). The latter direction of influence, i.e. the influence of trust or mistrust in elements of 'the system' on trust in its doctors, is pertinent to our project of assessing possible influences of personalized medicine—as the modus operandi of a healthcare system—on trust between patient and doctor. To wit, while trust grounded in the doctor's personality (e.g. his caring attitude, honesty, or high sense of responsibility for patients) should remain constant whatever the system's profile, system characteristics—in particular, in our case, the system's adoption of personalized medicine—can influence the doctor's perceived and even actual trustworthiness, as I discuss below.

It may sound prima facie odd that there should at all be any challenge to doctor–patient trust associated with personalized medicine, given that it is an advancement of medicine for the benefit of each individual patient. In this regard, two clarifications are in order. First, to the extent that personalized medicine is continuous with current medicine—e.g. if it does not proceed beyond a more precise sub-classification of diseases, based on genetic and other markers—then the caveats below will indeed be mostly irrelevant. Conversely, the more personalized medicine turns out to be a grand-scale big data revolution, the more it turns out to be a sort of 'paradigm shift' that ushers 'a fundamental change in the way medicine is practiced and delivered' (Pritchard et al., 2017, p. 143), the more serious the following challenges may become. The second clarification is that my intention is not to express scepticism but to explore potential challenges, of which we should be aware, so we can address them if necessary, in order to benefit optimally from the advantages offered by personalized medicine.

Researchers broadly agree that patients' trust in physicians consists mostly of the following dimensions: loyalty or caring, competency, honesty, and confidentiality (Hall, 2006). My discussion of personalized medicine refers to all these domains. I discuss: (i) patient's trust in the doctor's wisdom; (ii) trust as a function of a good rapport between doctor and patient (both (i) and (ii) are measures of the doctor's competency); (iii) trust related to confidentiality (here I will also touch upon honesty); and (iv) trust that the patient's wellbeing is the doctor's highest priority (a measure of caring).

Trust in the doctor's wisdom

Competence is among the most important factors responsible for trust in doctors. At times it is identified as the most important (Manderson and Warren, 2010). Although competence includes a number of capabilities, the most important, arguably, is clinical judgement. Accordingly, Skirbekk and colleagues interpret trust as 'the patient's implicit willingness to accept the physician's judgment in matters of concern to the patient' (Skirbekk et al., 2011, p. 1183). Good clinical judgement, in turn, comprises various parameters; two central ones that elicit trust are the doctor's relevant knowledge, and even more importantly—the doctor's wisdom. Now there is a reason to suspect that personalized medicine may exert negative influence on the dynamics of trust, as far as these parameters are concerned.

So far in medicine, the doctor has had knowledge of the natural course of the disease, which he could share with the patient. However, the more that personalized medicine becomes personal, the less will there be 'natural histories' of people's diseases, about which the doctor could counsel the patient. The reassuring effect of having a clear narrative to one's woes, which allows reasonable expectations, and the trust in the doctor who provides such a narrative, could diminish. More importantly, in medicine as we know it, doctors accumulate clinical experience and insight in treating patients suffering from specific diseases, which they can later share with new patients with the same disease. Clinical wisdom, based on years of practical experience, is an important capability through which doctors help patients plan strategies for coping with their illness, and optimize wellbeing. Beyond the citation of pieces of medical information, clinical wisdom expresses the 'art' facet of medicine. By 'wisdom' I mean understanding what is important when, including the insight into how to apply it successfully, and the commitment to do so. By 'clinical wisdom' I understand the integrative capacity to identify the salient considerations—both medical and para-medical—in each clinical case, and a good intuition regarding how to act accordingly to secure the best management and outcome for each patient.

What will happen to clinical wisdom if or when personalized medicine accomplishes its transformative potential? The more personalized medicine becomes personal, the more difficult it may become to apply wisdom in clinical practice. Medical recommendation will veer much more toward a machine-generated computational output. Such output will be the culmination of collecting genomic, transcriptomic, proteomic, epigenomic, metabolomic, and nutriomic data, as well as patient-related data on environmental, behavioural, and socio-economic factors. This data mountain will then be processed by algorithms of bioinformatics, biomathematics, and biostatistics for integration, analysis, and interpretation, and then to further algorithms to suggest optimal treatment (National Research Council, 2011). The doctor will understand little about the considerations that generated the computed medical recommendation. Indeed, often there will be nothing to understand, as the output will be the outcome of the brute force of mathematical and statistical computation. It will presumably also incorporate massive data generated without human interference via machine-to-machine communication through the Internet of Things. If this scenario or something resembling it comes to pass, the doctor might, at the limit, be reduced to a sophisticated clerk who communicates machine-generated bottom-line conclusions to the patient. No one knows precisely what the clinical encounter will look like under such circumstances, but there is reason to think that at least some aspects of clinical wisdom might be lost, and with it the more traditional sense of trust in the insight and good advice of a wise doctor.

Even today, there is a concern that doctors rely too heavily on medical decision-making algorithms, and that this has adverse effects on clinical judgement. In his book *How Doctors Think*, Jerome Groopman, Chair of Medicine at Harvard Medical School, asserts that 'algorithms discourage physicians from thinking independently and creatively' (Groopman, 2007, p. 5). In the revolutionary personalized medicine scenario, the shrinking-of-thinking problem can increase to qualitatively new levels. The doctor will no longer employ algorithms. He will more likely 'be employed' by them, having to construct much of his advice on parroting their output. To an extent, the doctor will become a 'Google searcher' in a world of big data medical informatics. It is worth mentioning that Groopman's insight is true more generally than in medicine: wisdom and following algorithms are two very different things. In fact, there is a sense in which they are *mutually exclusive*: whenever and to the extent that we can follow an algorithm, wisdom becomes superfluous. It is when there is *no* algorithm for how to think or act that wisdom reveals its indispensability. Indeed, we often need wisdom in order to choose the right algorithm for the situation at hand, but this is precisely because there is no meta-algorithm that governs the choice of the right algorithm for each situation. To the extent that trust is a response to wisdom, we should be vigilant lest a personalized medicine revolution diminishes trust in doctors.

Trust and the rapport between doctor and patient

Interpersonal trust thrives in conditions of intimacy and good communication. We are interested to examine the concern that a personalized medicine revolution might enhance some problematic trends in this regard.

While it would be silly to dispute the tremendous benefits that computer technology has brought into medical practice, a downside, which has received some attention, involves the effect of the introduction of electronic health records (EHRs) into the clinic on the communication pattern between doctor and patient. After a short greeting, doctors tend to spend the first minute or few minutes of the consultation interacting with the computer rather than interacting with the patient or discussing the patient's agenda (Duke et al., 2013; White and Danis, 2013). Often, it is the information on the screen that prompts the doctor's opening statement, instead of inviting patients to share their concerns and being responsive to them (Pearce et al., 2008). Rapport is often undermined even when physicians do finally attend to the patients themselves, as physicians type in data or gaze at the screen when talking to the patients or when the patients talk to them. According to one study (Margalit et al., 2006), physicians spend an average of 24–55% of the visit time gazing at the screen. Because consultation time is short and mostly fixed, and since physicians, like everyone else, have trouble multi-tasking—in the sense of interacting simultaneously with the computer screen and, in a meaningful way, with the patient (Booth et al., 2004)—time spent interacting with the computer is inversely related to physicians' focus on the patients themselves (Shachak and Reis, 2009). The result is that screen-driven communication inhibits patients' narratives and diminishes clinicians' responses to patients' cues about psychosocial issues and emotional concerns (Lown and Rodriguez, 2012). The rapport between doctor and patient is damaged. This state of affairs is anything but a marginal side-effect of the EHR era. Patient–doctor communication is a crucial component of the health-care visit, shown to have 'ramifications for patient satisfaction, adherence to treatment, conflict resolution, laboratory costs, and clinical outcomes' (Duke et al., 2013, p. 358).

Where face-to-face communication falters, the opportunity to forge relations of trust is diminished. Now in revolutionary personalized medicine, with its fundamental reliance on consulting big data edited via sophisticated bioinformatics, doctors' need for computer interaction to manage patients is likely to increase significantly. This is expected to add strain to good doctor–patient communication, and in turn to the sort of dynamics that forge trust.[1]

The challenge to attentive, face-to-face doctor–patient interaction is one aspect of the challenge of developing trust in the computer-age clinical encounter. A deeper issue is the very nature of a medical consultation in personalized medicine where technology becomes a central presence. Even in today's clinical practice, computers have been perceived as 'third parties' in the exam room, which compete with the patient for the clinician's attention (Lown and Rodriguez, 2012). The point is that trust flourishes in conditions of intimacy, and in this respect, the question is which elements that enter into the secure realm of the clinical encounter are experienced as 'external

---

[1] Topol et al. (2019) hope that personalized medicine and the use of information technology more generally will free time for longer doctor–patient interaction. I hope they are right. It is not clear that the tremendous advancements in medical technologies in, say, the past two generations had a net effect of freeing time for communication between doctor and patient.

agents' that intrude into doctor–patient intimacy. In his James Mackenzie Lecture, delivered at the Royal College of General Practitioners, Per Fugelli wrote: 'Modern times are characterised by the intrusion of external parties into the doctor–patient relationship. The autonomy of the general practitioner, and hence the capacity to create trust, is now compromised by big business, big government, big science, and Big Brother.' (Fugelli, 2001, p. 576). Perhaps, it is easiest to understand what is meant here by 'big business' being an 'external party' or a 'third party' or an 'external agent' in the doctor's office. The challenges posed by managed care to doctor–patient trust are a prominent example (Emanuel and Neveloff Dubler, 1995; Mechanic and Schlesinger, 1996). The idea that 'science' too can be an external agent that disturbs the intimacy of the clinical encounter may sound absurd. Perhaps, 'technology' would be the more appropriate term here. Fugelli writes of the examination room: 'In the Room of Trust the patient will feel secure with low technology and high fidelity. In the Room of Angst there will be cravings for multiple tests and sophisticated referrals.' (Fugelli 2001, p. 576). While this connection between trust and being satisfied with 'low-tech' medical wisdom does sound right with respect to classical medicine, the idea of technology as a 'third party' entering the clinical relationships is easier to understand in the world of personalized medicine, which, in its ambitious rendition, will rely on computers doing much of the medical thinking. Lee Hood (2019, p. 7) reports developing:

> a new type of artificial intelligence algorithm that has the ability to precisely reproduce the thinking of clinical experts for acute and chronic diseases . . . using virtual intelligent 'Lego blocks' which can interact with one another. The result is a deep reasoning system into which one can then feed the symptoms and data of an individual and which produces insights to help inform a physician in regards to diagnosis, therapy choice, and strategies for treatment in the future.

In this case it becomes easy to see how technology could be perceived, quite tangibly in fact, as an external agent in the exam room. Henrik Vogt (2017, p. 106) speaks ominously of advancing toward a 'computer–patient relationship'. Hence, even if personalized medicine manages to issue excellent advice, the erosion of intimacy will beget alienation. This might be a key challenge to the interface between personalized medicine as a producer of critical knowledge and the people whose life and wellbeing depend on this knowledge.

Trust and confidentiality

The intrinsic connection between confidentiality and trust needs no explanation; nor does the special importance of confidentiality in health matters demand elaboration. Indeed, confidentiality has long been a cornerstone of medical ethics. The meaning of confidentiality has been relatively straightforward in traditional medicine, with disputes being confined mostly to the question of the stringency of that duty, namely whether it is absolute or not, and if not, what should count as justified deviations (Gillon, 2001). The traditional idea of confidentiality in medicine has presupposed the paradigm of a private setting of doctor–patient interaction. Revolutionary personalized medicine, however, may bring significant change to the ethos and practice of confidentiality.

Because personalized medicine relies on genomic profiles, it is expected that the future standard of care will include whole-genome sequencing. This genomic (as well as other 'omics') profile will be added to central biobanks and large databases as part of clinical routine. In certain respects, this general scheme for patient management bursts the private setting paradigm of doctor–patient interaction. The nature of personalized medicine does not allow clear separation between patients as sources of data and patients as receivers of care; rather, in personalized medicine, patients' data must be mixed and processed with everybody's data. While privacy is the current paradigm and default of clinical care, sharing of data is a hallmark of personalized medicine and its conceptualization of patients' participation. This paradigm reversal would implicate the clinical encounter with problems of confidentiality associated with biobanks and large datasets—most notably, the problem that even anonymization does not exclude re-identification (Gymrek et al., 2013; Rodriguez et al., 2013; Craig, 2016; Harmanci and Gerstein, 2016). Any problem of confidentiality is *ipso facto* a threat to trust.

More generally, the dynamics of confidence surrounding information sharing and protection will be to an extent reversed in personalized medicine. Instead of personal medical information being situated first within the discreetness of the doctor–patient relationship, the first to know will be the computer and the people operating it. Those anonymous persons will have to determine the validity and significance of the information before they communicate it to the treating physician. Hence, in personalized medicine, the dynamic of entrusting one's private information with one's personal doctor—before any further decision of information sharing might be contemplated—will be reversed. This would omit a building block from the doctor–patient relationship that is conducive to forming relations of trust.

Challenges to trust due to problems of confidentiality can be aggravated by difficulties in finding the optimal model of informed consent. Since personal material will be incorporated into big data archives, either public or accessible by many users, established standards of informed consent may cease to be practical. I will mention three challenges: (i) Because personalized medicine collects great amounts of data regardless of any specific clinical need, nobody can tell in advance what information might surface. This can create difficulties for tailoring adequate informed consent. (ii) It will not be possible to tell patients what uses will be made of their personal information, since those uses cannot be foreseen. Yet people may care very much about their personal information not being used for causes to which they are ideologically opposed (Ewing et al., 2015). A notable example is public reluctance to allow commercial uses of genomic data (Caulfield and Murdoch, 2017). Hence the proposed solution of 'blanket consent', which is typically not an acceptable standard in modern medicine, will amount to a breach of confidentiality. (iii) In accordance with what was said above, it may well not be possible to tell patients exactly what risks to privacy they are undertaking. This may be further aggravated by a lack of adequate language for expressing categories of risk, given the apparent breakdown of the distinction between identifiability and non-identifiability regarding genomic data (Rodriguez et al., 2013).

The problem of informed consent for participation in biobanks is well known in research ethics, and poses a danger to public trust in researchers (Greely, 2007). In revolutionary personalized medicine, the problem will not only transfer to the clinic, but is likely to become deeper; for, while researchers can dismiss potential human subjects who would not consent, doctors cannot similarly simply dismiss patients. This may create pressure toward being less than clear and straightforward—or, more bluntly—less than completely honest about the details of privacy protection. Compromised standards of honesty are obvious threats to trust.

Trust and commitment to the best interests of the patient

A prominent ground for trusting is that the caregiver accords the highest priority to the patient's interests. The Medical Professionalism Charter posits altruism as a fundamental commitment of medicine, which contributes to trust (ABIM, 2002). That anyone should raise concerns regarding the priority of the patient's interests in personalized medicine may sound confused, given that personalized medicine's *raison d'être* is to provide tailored medical care to the individual patient. And yet, a potential challenge arises even in this regard. To understand the caveat, we should revisit the mechanism of medical decision-making in revolutionary big data personalized medicine.

Ambitious personalized medicine poses formidable challenges of data generation, management, integration, analysis, and interpretation. After having processed large-scale genomic data, there arise the challenges of interpreting the functional impacts of genomic variation, integrating data to relate complex interactions with phenotypes, and then translating these discoveries into medical practices (Fernald et al., 2011). Much of the challenge is of bioinformatics, namely, the hugely complex undertaking of reliably estimating many parameters and correctly inferring meaningful associations from multiple hypotheses tested simultaneously. (Attempting to reduce dimensionality in order to overcome difficulties of interpreting multiple testing involves the costs both of inserting measures of arbitrariness and of loss of information.) In short, omics data, let alone more holistic personal data, 'embody a large mixture of signals and errors, where our ability to identify novel associations comes at the cost of tolerating larger error thresholds in the context of big data' (Alyass et al., 2015, p. 10; Vogt et al., 2019). The upshot of all this is that while, *qua* personal, personalized medicine will have higher specificity for the patient, it could simultaneously face a problem of diminished accuracy. This may pose risks of harm to patients. Furthermore, we should emphasize in this respect that pattern recognition applied to big data by machine learning would not, by itself, provide causal explanatory basis for medical recommendations—unless overarching scientific theories for the conversion of multidimensional big data into mechanistic knowledge are developed and prove successful (Brenner, 2010). Without the relevant knowledge, medical doctors would be, in this scenario, at the mercy of machine-generated population-based correlations, with all their shortcomings. This, again, could risk harming patients, especially given the phenomenon of automation bias, in which 'clinicians unquestioningly accept machine advice, instead of maintaining vigilance or validating that advice'[2] (Coiera, 2018, p. 2331).

---

[2] It is true that the current alternative includes risks of decision-making bias; my point is that we should be vigilant to identify new risks, for only then can we opt responsibly for risk minimization in each relevant context.

We should hope that professional standards will successfully sift out results whose level of accuracy is insufficient; but the tricky question is what would count as accurate-enough? While it is reasonable to expect that personalized medicine will improve—perhaps very significantly—the healthcare of the average patient, or the average level of health, or put simply, will improve public health,[3] the ironic point is that this may not extend to every specific patient: the medical recommendation will be a function of wide-association studies, which we grant will be statistically responsible, but which nonetheless may not represent accurately the case of every patient. In conventional medicine too, the individual patient may suffer from treatment that was statistically established as helpful to the population overall, and this is precisely what personalized medicine sets out to correct; but in conventional medicine this dependence on population averages is the best that medicine can offer. In a revolutionary future of personalized medicine, on the other hand, we face the possibility that in certain cases the new algorithms will be much better for public health, because much better on average, but worse than conventional medicine for some individuals in some cases. This would drive a wedge between personalized medicine's value for public health versus for clinical medicine, in the sense of being the best treatment for each individual patient. Such a prospect can raise the *primum non nocere* objection in the context of personalized medicine. Protecting against breaches of *primum non nocere* is basic for medical ethics, and could have an adverse effect on trust in the healthcare system as well as in the doctor who is its representative.

Conclusion

It is indisputable that patients' trust in the healthcare system in general and in doctors in particular is of great importance for good healthcare. Trust is hard to win and easy to lose. Once trust is damaged, suspicion emerges; and relations marred by suspicion can in turn precipitate fast deterioration of trust. This is so both because it is of the inherent nature of mistrust that once its seeds are sown, it tends to grow quickly, and because mistrust tends to be reciprocated with defensive reactions—i.e. in the case of medicine, with defensive medicine—which further erode trust, setting in motion a down-spiralling vicious cycle. Hence, it is important to be vigilant in matters of

---

[3] Yet there are doubts regarding the optimality of personalized medicine even for public health. See Khoury and Galea (2016), Vogt (2017) and Wilson (ch. 13, this volume).

trust: to anticipate potential challenges and then, should they arise, to address them quickly and decisively. Medicine at its best should incorporate cutting-edge technology with optimal doctor–patient intersubjective dynamics. The tragedy of existence, however, is that ultimate harmonies are elusive—there always is some trade-off price to pay. The purpose of this chapter has been to identify potential causes of alienation and mistrust that might inflict the doctor–patient relationship the more cutting-edge technologies transform the intersubjective nature of the clinical encounter. Awareness of the potential challenges discussed in this paper increases the chances of optimizing the tremendous potential benefits of personalized medicine.

One of medicine's great future challenges will be to guard against big data, digital, genomic, and robotic medicine 'dehumanizing' the care relationship (Topol, 2019, p. 9). I will end with an apt quote from a recent paper in *The Lancet*: 'What does it mean to be a doctor? Is it still medicine we practice when a machine knows better than us our patient's diagnosis, treatment, or fate? Would the hand we hold at the bedside still be reassured by our words and care?' (Coiera, 2018, p. 2331). As this quote suggests, upholding the trust patients have always placed in doctors should be a guiding principle in the task of securing a humanistic future of personalized medicine.

13

When does precision matter? Personalized medicine from the perspective of public health

James Wilson

Introduction

Much recent health policy in the UK, USA, and elsewhere presupposes a tech-led vision of the future of medicine—as exemplified by the US NIH Precision Medicine Initiative, and the UK government's AI Early Diagnosis Mission.[1] This new approach aims to use large-scale datasets and rich information about individuals (particularly genomic or multi-omics datasets) to detect abnormalities that would have caused clinically relevant cases of disease, and enable early intervention. The underlying model of medicine is thus anticipatory rather than reactive: early intervention will ensure that a disease never eventuates, or that its effects are less severe. A lot of the excitement has coalesced around the model of P4 medicine: medicine which is predictive, preventive, personalized, and participatory (Hood and Friend, 2011; Vogt et al., 2016).

P4's focus on prevention is used to position it as superior when compared to previous models of clinical care.[2] However, the idea of prioritizing prevention is not new. Public health in particular has long focused on preventing disease and illness through interventions such as childhood vaccination or health and safety legislation. Within public health, primary, secondary, and

---

[1] For further on these policies, see Collins and Varmus (2015) and Her Majesty's Government (2019).

[2] See, for example, the following indicative passage: 'Surgery and other aspects of traditional medicine will be informed by the stratification of disease and of people where relevant. Focusing on the causes rather than the symptoms of disease will enable intervention to occur much earlier in the disease process, in many cases preventing disease from occurring in the first place.' (Flores et al., 2013, p. 567).

James Wilson, *When does precision matter? Personalized medicine from the perspective of public health*
In: *Can precision medicine be personal; Can personalized medicine be precise?*. Edited by: Yechiel Michael Barilan, Margherita Brusa, and Aaron Ciechanover, Oxford University Press. © Oxford University Press 2022.
DOI: 10.1093/oso/9780198863465.003.0013

tertiary prevention are standardly distinguished. Primary prevention aims to reduce the likelihood of disease or injury before it occurs (for example, anti-smoking or road safety campaigns). Secondary prevention aims to diagnose disease early in order to allow interventions that will minimize its effects (for example, cancer screening; training of employees to spot the signs of work-place stress). Tertiary prevention aims to minimize the effects of disease or injury that is already severe enough to have made a noticeable difference to the patient's life (e.g. rehabilitation programmes after a stroke; antiretroviral drugs after an HIV diagnosis).

Thus, what is distinctive in P4 medicine is not the idea of prevention per se, but highly targeted secondary prevention through the use of rich and large-scale datasets. Looking at P4 medicine within a broader account of preven-tion raises a question about whether the approach to prevention embodied in P4 medicine should be preferred to the well-established approaches to prevention embodied in classic public health. There is something conten-tious about P4 medicine's framing of secondary prevention as *early interven-tion*, where the contrastive 'later' intervention would be tertiary prevention. Secondary prevention counts as 'early' relative to tertiary prevention, but *late* relative to primary prevention.

If all that mattered was intervening as early as possible, then the 'early intervention' rationale for shifting from conventional medicine to P4 medicine would, by the same token, be an argument for shifting from P4 medicine to public health interventions that target primary prevention. What this shows is that if P4 medicine amounts to a superior model of medicine, it cannot be solely because it advocates acting early within a disease trajectory. It must be because it aims to do so in an appropriately *targeted* way.

The value of more precise targeting of interventions is usually taken for granted, rather than argued for explicitly; and it is assumed that the more tightly targeted an intervention, the better. The main argument of this chapter is that the value of precision targeting is more equivocal and con-textual than is often thought. Precision targeting of an intervention is im-portant only where there would be bad effects if the intervention misses its intended target—and this is far more likely to occur with some kinds of interventions than with others. Even where biomedical interventions are such that precision in their application would be valuable, the model of early intervention at an individual level on the basis of predictive modelling may struggle to give sufficiently accurate information. It is likely to lead to overtreatment, and to be less cost-effective and less equitable than public health approaches.

The overall conclusion is that ensuring that interventions have the optimal balance of benefit over harm is a much more important goal than precision, and precision medicine will tend to be a less effective way of pursuing the more important goal than public health-led approaches. In short, contrary to what is implied by policies such as the US NIH Precision Medicine Initiative, and the UK government's AI Early Diagnosis Mission, precision medicine could easily amount to a step backwards.

Precision

The more precisely targeted an intervention is, the more information those implementing it need to have about the targets at whom it is directed. Targeting interventions thus requires: (i) data collection, (ii) modelling or stratification and (iii) selecting or prioritizing an intervention based on risk profile. Depending on how precise the targeting is, more or less data, and a more or less sophisticated stratification model will be required. At the limit where an intervention is entirely untargeted, then little or no data collection will be required for targeting purposes. The degree of high dimensional variation in the phenotypic data collected (stage 1) will often be orders of magnitude greater than the variation in the kinds of interventions that can be applied (stage 3). This is one of the reasons why, after some initial enthusiasm for the terminology of 'personalized medicine', many shifted to talking of precision or stratified medicine, as it became clear that it is unlikely—at least in the short to medium term—that clinicians will be able to deliver on the promise of fully individualized therapies (Juengst et al., 2016).

For illustrative purposes, I shall use two formulations to contrast with precision targeting of interventions. First, *broadly targeted*, to suggest that an intervention is targeted on particular groups but not that tightly—for example where a diagnostic assay is run on a sample from a bacterial infection, and a suitable antibiotic chosen. Second, *untargeted* to suggest that an intervention is made available with minimal targeting at individuals—for example, a childhood vaccination campaign, in which it is recommended that all children receive the vaccine except the small number for whom this is biomedically contraindicated. Whereas the three categories of precisely targeted, broadly targeted, and untargeted will be useful for the purposes of exposition, it is important to be aware that there is a continuum from completely untargeted interventions to fully individualized interventions.[3]

Why think that precisely targeted interventions should be preferred to broadly targeted, or untargeted interventions? The value of greater precision is usually taken for granted, rather than argued for explicitly by proponents of P4 medicine. Where we do see arguments, they centre around *arguments from non-maleficence*, namely that untargeted interventions are harmful—or at least sub-optimal in the amount of benefit they provide—precisely because they are untargeted, and *arguments from waste*, namely that untargeted interventions are wasteful, just because they are applied to people who will not benefit from them.

Precision medicine is argued to be less harmful than untargeted approaches, because it will reduce 'spillover' effects on those individuals who would not have been selected to receive the intervention if it were very precisely targeted. Given the importance of the duty of non-maleficence, this might be taken to be a strong argument in favour of the importance of precision targeting of interventions. However, as the next sections explore, things are not so straightforward. First, it is wrong in theory to think that effective and ethically justifiable preventive activity requires precise targeting. Second, as currently constituted, P4 medicine is likely to generate models and interventions that are more precise than they are accurate—leading to overdiagnosis and overtreatment, and this itself raises questions about non-maleficence.

Are precision interventions less likely to be harmful?

P4 aims for a medicine that is predictive, preventive, personalized, and participatory. This might seem to suggest that there is an inherent connection between being personalized and being effectively preventative, i.e. that effective prevention is individually targeted. However, there is no intrinsic connection between prevention and personalization. While it is easy to think

---

[3] The distinction between precisely targeted, broadly targeted, and untargeted interventions describes the way that individual level data is used to tailor interventions, rather than the scale or richness of the datasets used to inform the intervention. It may be useful to make use of large and rich datasets even in broadly targeted or untargeted interventions. For example, post-marketing surveillance for pharmaceuticals may make use of rich data derived from a large number of individual patients. Insofar as this data is used to decide whether a drug should be withdrawn on safety grounds (or its indication narrowed), the resulting intervention would nonetheless be a broadly targeted or untargeted one.

of examples of health interventions that would be very harmful if they were rolled out in an untargeted fashion—such as chemotherapy, or brain surgery—there are also a range of health-related interventions where targeting is unimportant because the intervention would be either beneficial or at worst have no effect if it were made the subject of a blanket recommendation.

Alongside the rise of 'big data' early intervention approaches, recent years have also seen an increased emphasis on social prescribing for long-term conditions. That is, clinicians referring patients for health-system-recommended non-medical activities and support services, which might include exercise classes, weight management advice, visiting museums, joining a choir, or volunteering in a community gardening project (Drinkwater et al., 2019). Social prescribing embodies a very different approach to early intervention than is indicated by P4 medicine. Many of the interventions that fall under the category of social prescribing are non-specific. Clinicians could fruitfully suggest them to patients with a wide variety of conditions. Many of these are activities that individuals might want to do anyway regardless of whether they were ill—and so there seems to be little risk to patients in using such interventions in a broadly targeted or even an untargeted way. Some pharmaceutical interventions, such as statins, can also be prescribed safely in a broadly targeted way.

Knowledge dissemination and infectious disease control provide other cases of untargeted interventions that are widely thought to be good practice. Knowledge such as healthy eating advice is a non-rival good; if such information is provided freely to everyone, then many will be able to benefit without anyone losing anything. It is a feature rather than a bug of non-rival goods that they provide benefits to many beyond those at whom they were originally targeted (Wilson, 2012).

Goods such as herd immunity are network goods: the more who are vaccinated, the more beneficial it is for everyone. The greater the proportion who are vaccinated, the greater the overall benefit (and especially to those who would otherwise be at high risk). So in general it is a good rather than bad thing if a lot of people who would themselves be at low risk are vaccinated.[4] Careful targeting is ethically required only where activity undertaken for preventive purposes would create disproportionate risks of harm if undertaken

---

[4] Obviously, vaccine campaigns require at least a minimal form of targeting, to ensure those for whom they would be medically inappropriate, such as the immunocompromised, are not vaccinated.

in an indiscriminate way. Where an intervention's spillover effects are benefi-cial rather than harmful, then this is a good reason *not* to target it.

What implications do these thoughts have for thinking about precision medicine? One objection would be that there may be relatively few cases in clinical medicine where untargeted interventions are likely to be neutral or beneficial. Within clinical medicine it would be generally thought to count as a breach of nonmaleficence to provide an intervention which had no ef-fect, where an effective intervention is available.[5] There is something to this worry—but addressing it requires us to think more deeply about where and how in disease trajectories it is best to intervene, and how health systems should prioritize between different modalities of prevention.

Are untargeted interventions wasteful?

Untargeted interventions by their nature are undiscriminating about whom they affect, and so it is easy to assume that precisely targeted interventions will be more efficient. However, this does not follow. What counts as a wasteful (or an efficient) use of a resource should be defined relative to the overall goals of the health system. If the most efficient way of pursuing the goals of the health system involves applying an untargeted intervention, then such an intervention is by definition not wasteful.

It is true that there is a sense in which the intervention is 'wasted' if it is ap-plied to many who do not benefit from it. However, this sense of waste would be relevant for policy decision-making only if the main goal of the health system were to avoid applying interventions unless they could be shown to be biomedically beneficial for each individual to whom they are applied. Assuming without argument that this is a good goal for health systems would be question-begging.

Health system goals such as maximizing population health, and reducing health inequalities are often better served by untargeted interventions than by targeted interventions. (For a classic explanation of why, see Rose, 1981.) The World Health Organization-recommended standard of care for various neglected tropical diseases (NTDs), including schistosomiasis, is mass drug

---

[5] Hence the use of placebos in clinical trials has often been thought controversial, and it has been argued that there is an ethical requirement on doctors not to enrol patients in clinical trials unless they think that the trial intervention is equally as good as the standard intervention. (See Freedman (1987) for this view, and Miller and Brody (2003) for a critique.)

administration (MDA). MDA is explicitly untargeted, and has been defined as 'delivering safe and inexpensive essential medicines based on the principles of preventive chemotherapy, where populations or sub-populations are offered treatment without individual diagnosis' (Webster et al., 2014). MDA programmes can be *much* more cost-effective than testing individuals and providing treatment only where an individual tests positive. Indeed, Webster et al. (2014) argue that an integrated MDA programme targeting several different infections can be delivered for less than US$0.50 per person per year. Far from being wasteful, MDA is much more efficient than a targeted approach would be.

It is thus much too hasty to make the blanket assumption that precise interventions provide a better use of resources than untargeted ones. Whether precisely targeted interventions use available resources more efficiently than untargeted ones depends on what the goals of a health system are, alongside contextual features such as the cost of case-finding versus the cost of the intervention itself, and the cost per intervention of the untargeted as opposed to the targeted intervention.

These points have not always been well recognized in the precision medicine literature, where it has too often been presupposed that precision interventions will be more cost-effective than broadly targeted interventions—even where precision medicine involves bespoke pharmaceutical interventions being applied to the targeted individuals.[6] In reality, bespoke pharmaceutical interventions are unlikely to be anywhere near as cost-effective as loosely targeted pharmaceutical interventions using existing pharmaceuticals of much wider indication (Cossu et al., 2018). This is because of an obvious corollary of stratification and precision targeting of pharmaceuticals: the size of the target market will be significantly reduced—a shift that has sometimes been described as one from blockbusters to niche busters (Dolgin, 2010). Given that the size of target markets will be very significantly reduced, while the costs of drug development will not, the stratification and precision targeting of patented treatments would be expected to significantly raise drug prices (Gronde et al., 2017).[7]

---

[6] For example, Flores et al. (2013, p. 567) write 'Systems medicine will make disease care radically more cost effective by facilitating the stratification of both people and disease into distinct subgroups. Genomic analysis stratifies people into subgroups with different reactions to drugs, different disease risks and other clinically relevant factors . . . . These stratifications are providing increasingly more accurate diagnoses and cost-effective interventions based on the underlying causes of disease.'

[7] Where health systems take cost-effectiveness analysis seriously—as in the English National Health Service (NHS)—it is clear that introducing new drug therapies frequently leads to a net reduction of health gains within the health system. In a widely cited study, Claxton et al. (2015)

Secondary prevention, overtreatment, and overall benefit

Precision implies fine calibration and reproducibility, but it need not imply accuracy (Tal, 2017, section 8.3). A model that is the basis for interventions can be precise to an arbitrarily high degree, but yet be deeply unhelpful, if the model fails to provide a reliable basis for making interventions that will improve patients' lives. As Paul Kilanthi argued, there is good reason to think that it may be 'irresponsible . . . to be more precise than you can be accurate' (Kalanithi, 2014).

An ideal scenario would be to base interventions on measurements that are both precise and accurate, but this combination is difficult to achieve in the domain of clinical medicine. Whenever an intervention is targeted, there needs to be a test that matches individuals to interventions (or determines that no intervention is indicated). Tuning a test to minimize the chance of a type II error (a false negative), will tend to increase the chance of a type I error (a false positive), and vice versa. Thus, increasing the sensitivity of a test will typically also increase the chances of detecting cases that would not have led to clinically relevant disease—whether because they are false positives (misdiagnosis) or cases in which what is detected is genuinely incipient disease, but where this would not have gone on to cause a clinical problem (overdiagnosis).

Overdiagnosis can occur for a variety of reasons including that the disease would self-resolve, that it would have remained confined to a particular area (e.g. ductal carcinoma *in situ*), or that it is sufficiently slow progressing that the person would have died from another cause before it becomes problematic (Brodersen et al., 2014). Precision medicine, in so far as it aims to support targeted early intervention, is likely to make the problem of overdiagnosis much worse:

> monitoring many features of the body with highly sensitive technologies is bound to detect many abnormalities but without the ability to tell which,

calculated that the most likely value for the cost it takes to create one quality-adjusted life year (QALY) within the NHS is £12,936. However, when the National Institute for Health and Care Excellence (NICE) appraises new drugs to see if they should be recommended for use in the NHS, it rarely rejects pharmaceuticals that are priced at under £30,000 per QALY. The rather uncomfortable implication is that most of the drugs approved by NICE would be expected to displace larger health benefits than they create. Precision pharmaceuticals would, for the reasons given, be expected to be even more likely to lead to net health losses within the system than new non-precision interventions.

if any, will become clinically manifest. As a result, more people may be la-belled with more harmless conditions. (Vogt et al., 2019, p. 2)

In many cases the underlying processes of disease progression are sto-chastic. For example, it may be indeterminate whether a given very small cancer would have remained insignificant or if it would have become viru-lent if left untreated (Wu, 2020). It is thus difficult, if not impossible, even to determine whether overdiagnosis has occurred at an individual level. The extent of overdiagnosis can only really be explained and quantified at a popu-lation level (Hofmann, 2018).

P4 medicine, in other words, has a profile of risks that is much more similar to traditional secondary prevention policies such as breast cancer screening than its advocates sometimes realize. In both cases, the intervention targets individuals who do not consider themselves unwell. In both cases, despite the best of intentions, the policy may end up intervening in ways that do not make individuals' lives better, and crucially, it may not be straightforward for individuals to determine whether it is in their interests to accept the inter-vention. Due to this profile of risks, policies such as breast cancer screening have become much more controversial in recent years (Marmot et al., 2013).

In the case of cancer screening, the number needed to screen in order to save one life is in the thousands, so 'saving' one identifiable individual simultaneously means that there are thousands of individuals who are not benefited and who may be made worse off by the screening as a result of mis-diagnosis or overdiagnosis. Those whose early stage cancer is detected by screening and are successfully treated may feel that the screening has saved their lives, but no one is really in a position to say this, as in each case the treatment could have been overtreatment. So, on closer analysis, even the common assumption that screening *does* benefit identified individuals is contestable (Wu, 2020). The same worry would seem to apply to early inter-vention as promoted by P4 medicine.

Overtreatment occurs where an intervention would be expected to make the patient's life worse rather than better.[8] Whether a particular interven-tion counts as overtreatment would depend in part on the individual's values and preferences. For example, if an individual understands the relevant facts

[8] Overdiagnosis and overtreatment are closely aligned: to the extent that a disease is detected that would not have gone on to harm the patient, it is highly likely that an intervention that is applied to prevent the progress of that disease may make the patient's life overall worse—for ex-ample if it causes pain or other side-effects, or even if it is merely inconvenient.

and risk factors, but prefers to live with a particular health risk than to have the intervention, then giving this patient the intervention regardless would amount to overtreatment, even if other patients with a similar condition would prefer to receive the intervention.

Overtreatment is a potential problem for all kinds of medicine—but is a particular challenge in cases such as chronic disease, where interventions cannot provide a cure, but rather a way of helping to manage symptoms, prevent exacerbations, and improve quality of life. Some kinds of intervention will be more susceptible to giving rise to overtreatment than others. The profile of P4 medicine, like that of screening—in which interventions are applied to hundreds or thousands in order to prevent a single case, set up these conditions perfectly. Where an intervention provides a benefit that would also be valued by an individual quite separately from whether it would be necessary for improving their health, then it is much less likely that it will amount to overtreatment. Thus, interventions such as social prescribing are much less likely to lead to overtreatment than P4 medicine's early interventions.

Prioritizing public health, P4 medicine, and social prescribing

Much of the proposed benefit of P4 medicine comes from secondary prevention, and much of the benefit of public health comes from primary prevention. This raises an important question of how health systems should prioritize: which of the different potential preventive approaches should be given most weight?

I address these questions at much greater length in Wilson (2021), but we can sketch the following points here. Precision matters only where there is a likelihood of harm from an intervention being applied to those to whom it is not intended to be applied, so precision should not be a fundamental goal of health systems. However, goals such as cost-effectiveness, improving equity, and avoiding iatrogenic harm are all of core importance in health policy.[9] Given these goals it is plausible to think that some governments may

---

[9] I have argued elsewhere that primary prevention work through public health activity needs to be designed in a way that is ethically justifiable *given that* it is untargeted and that spillovers will occur. For example, any iatrogenic harms need to be justifiable *ex ante* to individuals who are harmed—perhaps explained by the benefits to health equity improvement in population health, or the overall prospects for representative individuals (Wilson, 2016). A similar principle will apply to precision medicine too: good early intervention requires finding an appropriate

currently be *over-valuing* P4 medicine relative to the benefits that other preventive modalities can bring.

A number of researchers have objected to this line of argument by claiming that it should not be a matter of either public health *or* P4 medicine, but *both*. On this view, supplanting public health is not the intention of P4 medicine, and so P4 medicine practitioners can agree that primary prevention through population-level public health measures is a superior bedrock for a health system strategy to secondary prevention through P4 medicine, without threatening their own position.

In response, I agree that primary and secondary prevention are compatible and complementary in ideal scenarios. For example, it is possible to support both tobacco control for primary prevention of cancer, as well as secondary prevention of cancer through personalized risk profiling and early intervention. However, primary and secondary prevention will often be in competition in practice due to limited resources—for example, additional resources for genomic sequencing and early intervention may come at the cost of reduced resources for smoking cessation.

Wherever there is population-level primary prevention activity that could be undertaken but is not as a result of cost (as will always be the case), and P4 medicine *is* funded to a significant degree, it is difficult to deny that the health system has implicitly prioritized between the two. Thus, I am unrepentant about the need to assess and prioritize P4 medicine against other preventive interventions including primary prevention, which compete for the same resources.[10]

Primary prevention through public health measures provides a more ethically defensible bedrock for a prevention strategy than P4 medicine for various reasons. It is simply easier to do more good more cost-effectively via primary prevention. Public health primary prevention is usually aimed at significant and common health problems, focusing on the social determinants of health, or reducing pollution, or operating childhood vaccination programmes. As such interventions operate to reduce or remove generic health risks, they do not usually need to be precisely targeted, as they will act to secure public goods and network goods, and to control generic risk

---

balance of the potential benefits of providing a diagnosis and getting a patient who benefits into treatment, against the potential harms of doing so.

[10] Beyond the practical incompatibility, there may also be disagreements between the literatures of public health and precision medicine—for example, about the extent to which genomics is relevant for disease causation and risk reduction (Chowkwanyun et al., 2018).

factors. Moreover, whole-population policies tend to be more equitable in their effects than ones that rely heavily on individual choice.

P4 medicine's interventions are much more precisely targeted than other interventions, but this does not suffice to give them a more optimal balance of benefit and harm. As we have seen, precision in detecting early signs of disease is compatible with (and may lead to) harm through overdiagnosis and overtreatment (Vogt et al., 2019). The underlying model of massive data collection in order to intervene in individual lives through secondary prevention is unlikely to be as cost-effective, or as beneficial for the worst off, as a model that emphasizes primary prevention through public health.

The discussion of social prescribing showed that there are some broadly targeted interventions, which provide individuals with access to a range of activities that are useful for improving wellbeing and resilience. These interventions are unlikely to make lives worse and are well placed to provide support for some of those with long-term conditions. So such interventions may score highly on desiderata of equity, and of avoiding iatrogenic harm, even though their current lack of a solid evidence base may make them weaker in terms of cost-effectiveness, and not an effective substitute for either large-scale public health, or for targeted clinical interventions.

In other words, P4 medicine may compete unsuccessfully with a range of other prevention modalities including public health and social prescribing, in so far as health systems prioritize cost-effectiveness, equity, and avoidance of iatrogenic harm. This conclusion should not be understood to be a claim of necessity, but rather a challenge for whether, and, if so, how P4 medicine could rethink itself to place more emphasis on these factors. For example, the cost-effectiveness of precision targeting could be greatly improved by focusing on how best to recombine or calibrate existing pharmaceuticals or other interventions that would be used widely in any case, or to select interventions that should not be performed. In such cases, so long as case-finding is reasonably cheap, it is plausible that precision approaches could be cost-saving relative to some less targeted approach.[11] Perhaps also, there will be some way in the medium term of much better distinguishing the signal from the noise when monitoring the vitals of millions for incipient risk factors.

---

[11] Thus, Nelson et al. (2019) provide a useful example of how a high-dimensional machine-learning model that predicts which patients are most likely fail to attend their appointments could save costs by prioritizing patients to receive telephone appointment reminders.

Conclusion

There are strong reasons to think that the availability of massively larger linked datasets should create a step change in the way that medicine is conceived, but it is too hasty to suppose that P4 medicine in its current incarnation provides the best future for medicine. Risk factors can be cost-effectively reduced for large numbers of individuals without individual targeting; and the fact that an intervention is untargeted and improves prospects for all can be an important part of what makes it equitable. If policy-makers are interested in creating a cost-effective and equitable health system that makes the best use of limited budgets, then P4 medicine as it is currently conceived looks like a questionable means of achieving this. Moreover, given the risks of overdiagnosis inherent in current iterations of P4 medicine, interventions such as social prescribing that aim to improve wellbeing through generalized interventions that are unlikely to cause harm may sometimes be preferable.

Much of this chapter has sought to deflate some of the over-optimistic estimations of what precision medicine can offer, but the aim of the critique has not been to provide in-principle objections to precision medicine. Rather, the aim has been to guide governments towards a more sober estimation of where the relative priorities should lie within the money they invest in prevention. Making perspicuous the ways in which primary, secondary, and tertiary prevention will in practice compete for funding should focus attention on the value for money each can offer. It is a dereliction of government's duty to divert funding to the new, high-tech solutions that will benefit relatively few, at the expense of getting the basics of primary prevention right.[12]

---

[12] If the genomics-led focus of P4 medicine is to be justified, this would need to be done in the light of the fact that it is likely for the foreseeable future to be significantly less cost-effective than some less targeted interventions. Perhaps what Norman Daniels (2007) has sometimes described as the 'specialness' of healthcare justifies spending much more per QALY on precisely targeted healthcare interventions than on public health interventions, for instance. For what is probably the best current attempt to make this case, see Badano (2016). I criticize such approaches in Wilson (2009).

14

CRISPR—a challenge for national and international regulation

Dianne Nicol

Introduction

In this chapter I bring my perspective as a law academic to a consideration of the extent to which modern medicine can realize its full potential through greater precision and personalization. This perspective is influenced by my immersion for the past 25 years in the science of genetics and genomics, and in the ethical and social milieu within which the science is positioned. The chapter focuses on the role of law and other forms of regulation in this new precision medicine era (which I see as embracing diagnoses, therapies, and treatments targeted to the needs of individual patients based on their genetic and other predispositions: Ramaswami et al., 2018). In limiting my discussion in this way, I acknowledge that there are other forms of personalized healthcare, including body imaging, telehealth and the like, each of which raises its own distinct ethical, legal, and social concerns (Nuffield Council on Bioethics, 2010). However, the close connectedness between precision medicine and genetic predisposition is of particular interest to me in this chapter, not least because it is widely acknowledged that there is a failure to reflect genetic diversity in the development of precision medicines, which can have a profound impact on distributive justice (Cohn et al., 2017).

The challenge is to provide an appropriate regulatory framework to facilitate new technological developments, but only in ways that recognize and respect the rights and freedoms of individuals and the broader values of society as a whole. This will require a regulatory system that not only promotes safe, efficient, and effective access to the technology, but also reflects ethical norms and societal attitudes towards that technology.[1] The regulatory system

---

[1] Although it may seem obvious to state that the law should reflect ethical norms and social values, it is recognized herein that this point is not uncontroversial, particularly in the legal philosophy literature. From the Hart–Dworkin debates in the mid-20th century to the present day,

Dianne Nicol, *CRISPR—a challenge for national and international regulation* In: *Can precision medicine be personal; Can personalized medicine be precise?*. Edited by: Yechiel Michael Barilan, Margherita Brusa, and Aaron Ciechanover, Oxford University Press © Oxford University Press 2022.
DOI: 10.1093/oso/9780198863465.003.0014

must regulate the development, approval and use of precision medicine in ways that are clinically robust—safe and efficient, and for patient benefit—but also socially robust—accountable, democratic, and trustworthy (Nicol et al., 2017, p. 88). This means that, from the perspective of the individual, due attention must be given to such matters as freedom to choose whether and to what extent to participate, protection of privacy, and freedom from discrimination. Social accountability further demands that due attention be given to such matters as distributive justice, solidarity, and benefit sharing. These are big asks for any regulatory system, particularly so when decisions have to be made about what should be allowed, what should be prohibited, and what should be prioritized.

I use genome editing as a case study with which to explore these issues. In using this term, I intend to include all techniques that cause site-specific double-strand breaks in the DNA molecule (Gaj et al., 2013). I should note at the outset that use of the term 'editing' is not uncontroversial. For instance, in a study analysing the use of metaphors in this field, Meghan O'Keefe and colleagues conclude that metaphors such as 'editing' 'inaccurately convey precision on the one hand and obscure what is not currently known about [the technology] on the other' (O'Keefe et al., 2015). These reservations about the use of such metaphors are not trivial, particularly because of the ways in which such language choices can 'drive and shape issue framing and policy content' (O'Keefe et al., 2015).

Recognizing the force of these reservations about the use of language, I have nevertheless chosen genome editing technology for analysis in this chapter (and chosen, for simplicity, to continue to use this language of genome editing) largely because it has the potential to be the most highly personalized form of medicine, in the sense that it involves modifications to our own individual genomes, what many people perceive as touching the essence of personhood. It is also precise, in that it involves very specific modifications to discrete elements of these genomes. Genomic modifications are not just capable of being undertaken on consenting adults. They can also be applied to human gametes, embryos, foetuses, and infants. Given that these interventions can affect both future people and existing people who currently lack capacity to consent, we must ask, as a society, whether this is a path we wish to follow. This chapter illustrates that, irrespective of whether the focus

legal philosophers have wrestled with the relationship between legality and morality (see especially: Hart, 1977; Dworkin, 1977). For a useful appraisal of the key elements of the debate see Shapiro (2007).

is on heritable or non-heritable forms of genome editing, the regulatory challenges are significant.

Genome editing—a rapidly advancing form of precision medicine

Technological developments in genome editing

Genome editing is widely seen as having the real prospect of improving the health and wellbeing of individuals around the globe (Turitz Cox et al., 2015). The discovery in 2012 that naturally occurring Clustered Regularly Interspaced Short Palindromic Repeats (CRISPR) and CRISPR-associated proteins in prokaryotic cells could be modified and utilized for editing the genome in other species has been particularly influential, both in basic and applied research involving manipulation of the genome and in the translation of these scientific advancements into practice (Maeder and Gersbach, 2016). There are many useful texts and web-based resources that explain the scientific aspects of CRISPR and other genome-editing technologies. In essence, CRISPR involves making precisely guided double-stranded cuts in the DNA molecule using constructs combining RNA (the guiding component) and enzymes (the cutting component). The cuts can then be left to repair through natural processes, or can be guided in various ways to more directed repair pathways (Ye et al., 2018).

CRISPR and other forms of genome editing are already transforming agriculture by providing more flexible, reliable, and simple approaches than traditional techniques for modifying genetic traits (Gao, 2018). When combined with 'gene drives', genome editing has the potential to eradicate vector-borne diseases like malaria and dengue fever (Champer et al., 2016). Each application carries with it a raft of ethical, legal, and social complications, which are beyond the scope of this chapter to fully examine. In the context of precision medicine, CRISPR and other genome-editing tools may be used in the future to treat a large number of diseases associated with mutations in somatic cells. According to clinicaltrials.gov, 35 somatic cell genome-editing clinical trials were already in progress at the start of 2020 (Eckstein and Nicol, 2020). However, like more traditional clinical trials of any kind, many of these will fail during the clinical trials process, and it will likely be many years before the remainder receive marketing approval (Ginn et al., 2018).

It had been thought that genome editing tools might also be employed some time in the future to make germline alterations in human gametes or embryos that would be passed to subsequent generations. Until recently there appeared to be international consensus that it would be inappropriate to do so until more research had been undertaken and communities around the globe had been given the opportunity to deliberate on the appropriateness of such interventions (Baltimore et al., 2015). However, in November 2018 an announcement was made that a scientist (Dr He Jianku) had modified the genomes of a number of human embryos using CRISPR and had implanted two sibling CRISPR-modified embryos into their mother, who had subsequently given birth (Cyranoski, 2019a). It has been widely reported that He Jianku's aim was to disable the CCR5 gene in order to mimic a natural mutation that has been found to confer resistance to HIV (Reardon, 2019). His announcement led to swift condemnation across almost all of the scientific community, in part because of the questionable effectiveness of the intervention, the risks associated with off-target effects, and the risk that there could be disruption of other functions of the CCR5 gene. It also raised serious questions about the ethical appropriateness of the intervention, and led to a call for a global moratorium on the use of this therapy in the context of the germline (Lander et al., 2019).

Policy advances in genome editing

The policy debate has advanced since Dr He made his announcement. One important step in this process has been the establishment of an Expert Advisory Committee on Developing Global Standards for Governance and Oversight of Human Gene Editing by the World Health Organization (WHO). Although the Committee has the mandate to consider all aspects of genome editing in humans, its first step was to develop specific recommendations for germline editing. In July 2019, the WHO Director General accepted a recommendation of the Committee that regulatory authorities in all countries should not allow any further work in this area until its implications have been properly considered (World Health Organization, 2019). Since then, the Committee has focused more attention on somatic cell editing, including the launch of a registry to record genome-editing clinical trials and an Internet consultation process (Hamburg and Cameron, 2019). In July 2021, the Committee released a set of three documents putting forward a governance framework for human genome editing (World Health Organization, 2021a, 2021b, 2021c).[2]

Another international body has also been created to examine clinical applications of germline editing, the International Commission on Clinical Use of Heritable Human Genome Editing (ICHHGE). The Commission was convened by the US National Academy of Medicine and National Academy of Sciences and the UK Royal Society.[3] The Commission released the report on its inquiry into clinical use of heritable human genome editing in September 2020 (International Commission on Clinical Use of Heritable Human Genome Editing 2020). As I have noted elsewhere,

> Central to the report's recommendations is a cautious 'amber' light (for germline genome editing) at some future time—but only in situations where (i) it is not otherwise possible to create genetically related offspring free from serious monogenic conditions; (ii) a set of onerous criteria are satisfied; and (iii) there are assurances of safety, efficacy, and efficiency. Though some might argue this is effectively a 'red' light, such a precautionary approach is entirely appropriate, given the issues at stake (Angrist et al., 2020, p. 339).

Although these international policy developments are valuable, they do not have the force of law to prevent the birth of another so-called 'CRISPR baby' somewhere in the world at some time in the future. Nationally enforceable laws are required if such activities are to be prohibited. Yet the science of genome editing marches on, and it is increasingly recognized that the science is outpacing regulatory responses. Jennifer Doudna, one of the pioneers of the CRISPR technique, encapsulates the concerns that such developments should be raising for us all, noting that: '. . . I believe that moratoria are no longer strong enough countermeasures and instead, stakeholders must engage in thoughtfully crafting regulations of the technology without stifling it' (Doudna, 2019, p. 777). The ICHHGE (2020) recognized this need for national regulation in its report, specifically recommending that competent

---

[2] World Health Organisation. 2021a. *Human Genome Editing: Position Paper.* Available at: https://www.who.int/publications/i/item/9789240030404; World Health Organisation. 2021b. *Human Genome Editing: Recommendations.* Available at: https://www.who.int/publications/i/item/9789240030381; World Health Organisation. 2021c. *Human Genome Editing: A Framework for Governance.* Available at: https://www.who.int/publications/i/item/9789240030060

[3] Details on the work of the Commission are available at: http://nas.edu/gene-editing/international-commission/index.htm

regulatory bodies must be in place in any country considering clinical use of heritable human genome editing (Recommendation 8).

In parallel with this work on germline, or heritable genome editing, we are seeing less focused discussion on the appropriateness and adequacy of current regulation of somatic cell genome editing, both in academic commentary and international policy debates. However, I argue that somatic cell genome editing raises particular concerns, in part because it is a novel technique and the risks associated with manipulating the human genome using this technique have not yet been well tested, but also because of the questions it raises about distribution of healthcare resources and allocation of healthcare budgets. Admittedly, the idea of therapeutically manipulating the human genome for health-related purposes is not new. As early as the 1970s scientists began exploring the potentialities of recombinant DNA technology through early stage experimentation. Scientific concern about the potentially adverse consequences of using this powerful tool to manipulate the genome were also recognized (Berg, 2008). By 1976, the US National Academy of Science had established the Committee on Recombinant DNA to act as an oversight body. By the 1990s, human gene therapy clinical trials had begun (Cohen-Haguenauer, 1995). However, the tragic death of a participant, Jesse Gelsinger, illustrated the risks associated with early stage implementation of such techniques, and the challenges involved in ensuring that there are no cracks in the regulation and ethical oversight of such trials (Savulescu, 2001). He received gene therapy to treat a genetic disease resulting in a failure to metabolize nitrogen, with the cause of his death alleged to be an immune response to the viral vector (Savulescu, 2001). There is a body of commentary that alleges a series of failures in the ethical review processes relating to the treatment received by Jesse Gelsinger, as well as financial conflicts of interest (Shrebnivas, 2000; Savulescu, 2001).

A need for increased focus on distributive justice

Although the quantum of gene therapy clinical trials slowed for a time following the Gelsinger tragedy, the promise that therapeutic interventions could repair defects in the human genome (as well as genomes of other species) led to further research, which is now starting to reap rewards (Ledford, 2019). However, high costs associated with bringing treatments to market, small market size, and high number of failures inevitably lead to staggeringly expensive treatment options in order that proponents have the opportunity to secure cost-recovery. For instance, in May 2019, approval was given by the US Food and Drugs Administration for Zolgensma, a gene therapy for the treatment of spinal muscular ataxia, marketed by Novartis at a cost of over US$2.1 million for a one-off treatment, making it the most expensive therapy on the market (Anonymous, 2019a). Other emergent gene therapies carry only slightly smaller price tags, making them inaccessible without government or philanthropic support or insurance cover. The tragedy for individuals from some of the poorest parts of the world who suffer from diseases such as sickle cell anaemia is that long-awaited therapeutic advances are priced well beyond their reach (Anonymous, 2019b; Ledford, 2019). The consequence is that the ideals of personalization and precision in delivery of treatments for some of the most pernicious diseases globally could be lost in the mire of economics for many of the world's population, potentially sidelining any attempt to meet the noble bioethical goals of distributive justice.

As Leslie Francis and others have noted, justice comes fourth in the list of widely accepted bioethical principles, behind autonomy, beneficence, and non-maleficence, which may suggest that it is somehow seen as less important (Francis, 2017; Reynolds, 2020). Yet as a special Hastings Center Report rightly acknowledges (Reynolds, 2020, S3), justice should actually take centre stage in the use of genomic knowledge in medicine:

> . . . the conviction and hope that justice is at the normative heart of medicine and that it is the perpetual task of bioethics to bring concerns of justice to bear on medical practice. On such an account, justice is medicine's life-blood, that by which it contributes to life as opposed to diminishing it.

Distributive justice is failing to find traction when it comes to genome editing research and clinical trials, even in regions like Europe and North America. Individuals of European ancestry are often favoured in clinical trial recruitment, risking skewing results and potentially leading to uneven distribution of improvements in diagnosis and treatment (Cohn et al., 2017). Ethnically and socially disadvantaged groups, scarred by past experiences, are reluctant to participate in research and clinical trials (Hildebrandt and Marron, 2018). Discussions relating to disabled people tend to focus on their medical conditions rather than their social situations (Wolbring and Diep, 2016). The benefits arising from these research efforts are even less likely to be available to the vast majority of the global population in the developing world, as observed from the broader experience with access to medicines (Kiddell-Monroe, 2014).

The regulatory hurdles faced by proponents of innovative somatic cell genome editing and other precision medicines doubtless contribute to the costs of product delivery, which are inevitably passed on to end-consumers. Yet there seems little appetite to closely scrutinize this regulatory milieu at the global policy level. A 2017 report by a US-based committee on the science, ethics and governance of human genome editing did focus some attention on the regulation of somatic cell genome editing (National Academy of Sciences, National Academy of Medicine, 2017). The committee concluded that (p. 6):

> Overall . . . the ethical norms and regulatory regimes developed for human clinical research, gene transfer research, and existing somatic cell therapy are appropriate for the management of new somatic genome-editing applications aimed at treating or preventing disease and disability.

I argue that there continues to be a range of scientific, social, legal and ethical questions with which policy-makers and legislators must grapple in the context of somatic cell genome editing as well as germline editing. As I and my colleagues have noted elsewhere, there are still a number of key areas of regulatory confusion, requiring responses that are accountable, democratic, and trustworthy (Nicol et al., 2017). Not least of these is how to achieve a balance between facilitating innovation and ensuring equity of access.

The legal and regulatory milieu for genome editing

Throughout this chapter I have repeatedly used the language of 'laws and other forms of regulation'. The rationale for this expansive language is because there are many ways by which precision medicine activities are controlled and managed, beyond traditional statutory instruments. Scholars in regulatory theory, including Julia Black and others, provide useful guidance on the scope of the concept of regulation (see, e.g., Black, 2001, 2014). In my work with my colleagues on the regulation of innovative health technologies, we have adopted the following broad definition formulated by Black: 'regulation, or regulatory governance, is the organized attempt to manage risks or behaviour in order to achieve a publicly stated objective or set of objectives' (Black, 2014, p. 4). This definition brings a wide array of matters within its scope: 'formal, direct mechanisms—e.g. statutes, regulations, policies—as well as less formal, and less direct mechanisms—e.g. industry self-regulation, stakeholder forums, funding decisions' (Kaldor et al., 2020, p. 287).

One problem in this area is the sheer volume of laws and other forms of regulation that impact on the progress of genome editing and other forms of precision medicine as they move from the research laboratory to more mainstream clinical practice. Included in the body of laws and other regulatory instruments applicable to precision medicine in general, and genome editing in particular, are generalist consumer protection, privacy, anti-discrimination and tort laws, specific health- and technology-related laws (such as prohibitions on reproductive cloning and regulation of research involving human embryos, regulatory approvals for drugs and diagnostics, and regulatory requirements for genetic modifications), intellectual property laws, internationally recognized good practice standards, professional guidelines and more (Sherkow, 2019). With a diverse group of colleagues, I have described this complex and intersecting body of laws as a 'regulatory soup' (Nicol et al., 2016). We make the argument that there is a need for more socio-legal research in this area, to assess whether it is possible to ease the path through this regulatory soup and to better ensure that the regulatory requirements for precision medicine are efficient, effective, and transparent.

Prohibiting unacceptable practices

There will be some instances where, no matter how innovative a technology might be, it will not be acceptable to contemporary society. Many societies have chosen, through their legislators, to absolutely prohibit reproductive cloning. In my own country, Australia, for example, reproductive cloning is prohibited through the *Prohibition of Human Cloning for Reproduction Act 2002* (Cth), with a maximum penalty of 15 years imprisonment. When enacting this legislation, the government of the day chose to add other practices that they deemed inappropriate in the context of human reproduction. Relevant prohibitions in sections 12–15 of the Act include (but are not limited to):

- intentionally creating a human embryo by a process of the fertilization of a human egg by a human sperm outside the body of a woman, other than to achieve a pregnancy;

- intentionally creating a human embryo containing genetic material pro-vided by more than two persons by a process of the fertilization of a human egg by a human sperm outside the body of a woman;
- intentionally developing a human embryo outside the body of a woman for a period of more than 14 days, excluding any period when development is suspended; and
- altering the genome of a human cell in such a way that the alteration is her-itable by descendants of the human whose cell was altered; and it was in-tended that the alteration be heritable.

One consequence of this prohibitory language is that it captures technological advances that were not in contemplation when the legisla-tion was enacted, including both genome editing and other techniques. One example is the technique of replacing mitochondria that bear genes carrying deleterious mutations with mitochondria that do not carry these mutations. An explanation of the nature of this form of therapeutic inter-vention is probably needed at this point. Briefly, mitochondria are found in all cells in the human body and are the sites where cellular respiration occurs. They carry their own sets of genes, separate from the nuclear genome. Mutations in mitochondrial genes may cause mitochondrial disease, symptoms of which range from lack of energy to early death. Mitochondrial donation is a form of reproductive technology whereby an embryo is created using the nuclear DNA from a man and a woman, and the mitochondrial DNA from an egg donated by another woman (National Health and Medical Research Council, 2020a, pp. 12–13). To date, the only country that has specifically legislated to allow mitochon-drial donation is the UK, albeit under a strict regulatory regime. Public deliberations on the legitimacy of this type of intervention are occurring in other jurisdictions, including Australia (National Health and Medical Research Council, 2020b).

Although scientifically distinct from genome editing, there is a risk that mitochondrial donation could not only be captured by old laws designed to prohibit particular activities (in this case reproductive cloning) but could also become entangled in the debate about germline genome editing. This is not meant to suggest that we should shy away from enacting laws that prohibit certain scientific activities, but to recognize that laws, of their nature, tend to be blunt and inflexible instruments. For this reason, it is vital that mechan-isms are built into those laws that apply in areas of rapid technological change to allow for them to be regularly revisited and updated, to ensure they continue to reflect societal values and ethical norms.[4]

In the context of genome editing, the impact of the *Prohibition of Human Cloning for Reproduction Act 2002* (Cth) is that germline editing is prohibited in Australia (Nicol, 2020). Moreover, it has been suggested that this statute, in combination with its sister statute, the *Research involving Human Embryos Act 2002* (Cth), creates a regulatory environment within which research involving the use of CRISPR technology in human embryos could not be undertaken in Australia (Taylor-Sands and Gyngell, 2018). This second piece of legislation creates a licensing regime for use of human embryos for research, with unlicensed activities being prohibited. Use of embryos for research involving CRISPR may not be a licensable activity under this legislation (Nicol, 2020).

Australia is not alone in imposing a restrictive regulatory framework on germline genome editing and human embryo research. Germany and Canada, to name but two countries, have similarly strict regimes (Faltus, 2020; Kleiderman, 2020). Other countries have a somewhat more open research environment, though most require some form of regulatory oversight for any research involving human embryos. Though not all countries have legislation creating criminal offences for germline genome editing, many have either legislation or another form of regulation that prohibits/deters/discourages germline editing and provides oversight of human embryo research (Boggio et al., 2020). The concern that is probably most alarming to Jennifer Doudna and others is the absence of such regulatory oversight in some other countries (Doudna, 2019), which, coupled with weak enforcement measures in others, and lack of consistency in approaches globally, is hardly desirable. Essentially, it establishes the preconditions for an unscrupulous and unregulated germline genome editing tourism market to emerge, preying on the desire of would-be parents for personalized interventions that might allow them to have biologically related offspring. Recent statements by a Russian scientists of a desire to create more CRISPR babies fuels these concerns (Cyranoski, 2019b). Arguably, making this technology available in a well-regulated domestic environment may be a better option than this

---

[4] By way of example, the Australian *Prohibition of Human Cloning for Reproduction Act 2002* (Cth) and *Research involving Human Embryos Act 2002* (Cth) included two mandatory reviews, five years after the legislation first entered into force and three years after the first review. However, no further mandatory reviews were included in the legislation (see Nicol, 2020).

dystopian alternative. But the step of creating a regulatory regime that allows germline editing is so significant that it should only be considered if there is broad societal and scientific consensus to do so.

Regulating more acceptable practices

Unlike germline genome editing, the question whether somatic cell editing should be allowed in some form or other is under debate. Nevertheless, there remain many unanswered and important questions about specific aspects of the regulation of this form of genome editing. Without attempting to be comprehensive, the following series of questions encapsulate some points of contention.

- Should somatic cell genome editing be regulated as a therapy or a device; as a product or a process? Answers to these questions have significant implications in terms of the regulatory pathways that will need to be followed. The regulatory pathways for drugs and biologic products, on the one part, and medical devices, on the other, tend to be discretely siloed in many jurisdictions. Furthermore, therapeutic products tend to be much more strictly regulated than therapeutic processes or methods (Nicol et al., 2017).
- How should efficacy and safety of personalized genome editing constructs be validated in a clinical trials regime that is based very much on large-scale randomization? Although this is not, strictly speaking, a legal question, the regulatory framework for clinical trials is designed in such a way as to accommodate large-scale randomized trials and any changes in approach will require amendments to this framework.
- Should the regulatory regimes that have been created in many countries for licensing and monitoring of the release of genetically modified organisms play a role in the regulation of somatic genome editing? This is a live issue, with differing approaches between countries. In Australia, for example, it would appear that the *Gene Technology Act 2000* (Cth) only applies to the genetically modified viral vectors designed to deliver CRISPR constructs into human cells. The Act appears not to apply to the constructs themselves, to the modified human cells, or to forms of delivery other than through viral vectors (Eckstein and Nicol, 2020).

- Who should pay for somatic genome editing (Kozubek, 2017)? In answering this question, governments will need to reflect on such issues as whether to prioritize genome editing as against other healthcare needs in setting remuneration schedules. Although such issues are essentially economic, solutions will need to be implemented through regulatory frameworks (Touchot and Flume, 2015; Marsden et al., 2017).

- What, if any, limitations should be placed on the exercise of intellectual property rights in this context? Screeds have been written about the role of patents in limiting access to drugs to the poorest members of the global community, and the extent to which globally agreed flexibilities in patent laws can ameliorate some of these limitations (World Health Organization, World Intellectual Property Organization and World Trade Organization, 2013). There is likewise a large body of literature discussing the extent to which patents over foundational genetic technologies are prone to slow the pace of biomedical innovation. Michael Heller and Rebecca Eisenberg's 'anticommons theory' in 1998 was one of the major catalysts for a body academic research aimed at understanding how gene and related patents impact on biomedical innovation (Heller and Eisenberg, 1998). In response to the concerns raised by Heller and Eisenberg and others, in 2007 the Organisation for Economic Co-operation and Development (OECD) called for member nations to adopt patent licensing practices that:

> should foster innovation in the development of new genetic inventions related to human healthcare and should ensure that therapeutics, diagnostics and other products and services employing genetic inventions are made readily available on a reasonable basis (OECD, 2007, Principle 1A).

In the field of genome editing, we are already seeing a rush to patent and commercialize foundational elements of the technology (Egelie et al., 2016), and an exploration of the legal and ethical consequences (Sherkow, 2017; Feeney et al., 2018). For this technology to fulfil its promise to society as a whole, nations and patent holders will need to fully commit to the aspirations of the WHO, the OECD and other international agencies to facilitate access.

The issues raised in these last two dot points, in particular, exemplify the challenge in addressing such fundamental questions as how should genome editing be prioritized in an overwhelmed global healthcare budget, and how might the benefits of genome editing be distributed equitably?

The influence of public views on setting the regulatory environment for genome editing

The work of the WHO Expert Advisory Committee on Developing Global Standards for Governance and Oversight of Human Gene Editing in developing a governance framework for human genome editing provides some high-level assistance in addressing some of the legal issues raised above (World Health Organisation, 2021a, 2021b, 2021c). The International Commission on Clinical Use of Heritable Human Genome Editing (2020) also provides useful guidance in the germline context. Both of these bodies recognize the importance of community engagement in this process.

The assessment of societal responses to new technological developments has both normative and practical aspects. The traditional ethical principles of autonomy, beneficence, non-maleficence and justice, together with other emerging ethical principles such as solidarity and benefit sharing, are vital touchstones in determining what is acceptable to society. We also need to understand more practically how contemporary society views technology: what is viewed as acceptable use and what is not. In areas of rapidly developing technology, in particular, policy-makers and legislators must be guided both by ethical norms and by views of society towards new technologies, not just by expert opinions.

Research has shown that people in democratic countries have confidence in science, particularly biomedical science, with consistent findings that Australians, Europeans, and North Americans are generally positive about genomic research if it is directed towards discoveries that have beneficial applications that are in their interest (Bruce and Critchley, 2008–2017). In the specific context of genome editing, research is showing that people have clear preferences for progressing somatic cell genome editing over germline editing, although they do not necessarily oppose germline editing outright. Rather, the clearest distinction they draw is between therapeutic and enhancement interventions (McCaughey et al., 2016; Gaskell et al., 2017; Rose et al., 2017; Scheufele et al., 2017; Weisberg et al., 2017). This point is illustrated in my own research with colleagues from a range of disciplines, involving an Australia-wide survey of attitudes towards germline gene editing (Critchley et al., 2019). The results of the survey of more than 1,000 Australians suggests that there is some level of comfort with editing human and animal embryos for research purposes. However, the survey also revealed that many respondents did have ongoing concerns about the consequences of germline genome editing.

Calls for deeper engagement with members of the public are gaining traction. For example, in 2016 the international Association for Responsible Research and Innovation in Genome Editing (ARRIGE)[5] was created, with one of its principal aims being to bring civil society into the discussion about responsible regulation of genome editing (Chneiweiss et al., 2018). Sheila Jasanoff and Benjamin Hurlbut have further proposed a global observatory on genome editing to provide for a 'cosmopolitan conversation' on genome editing (Jasanoff and Hurlbut, 2018). In the specific context of germline genome editing, the International Commission on Clinical Use of Heritable Human Genome Editing (2020, Recommendation 2) has recommended that '[e]xtensive societal dialogue should be undertaken before a country makes a decision on whether to permit clinical use'. However, we still await more fully rounded public deliberation on the complex issues associated with genome editing. Together with a broad group of interdisciplinary scholars, I and my colleagues at the Centre for Deliberative Democracy and Global Governance at the University of Canberra have proposed an ambitious plan for a global genome editing citizen deliberation as a means of providing one vehicle for such deliberations (Dryzek et al., 2020).[6] We are not suggesting that the outcomes of such deliberations could possibly give policy-makers definitive answers to questions about the paths they should take in prohibiting, permitting, regulating, and incentivizing the various aspects of genome editing. Ethical norms, expert opinions, economics and other factors remain relevant. What we are suggesting is that the voices of ordinary citizens must also be heard, and we need to develop robust methodologies to achieve this end.

Conclusion

As the calls for a voluntary moratorium on germline genome editing continue, we need to reflect as a global society on what exactly a moratorium might achieve, whether it is appropriate, and whether it is likely to be effective. Ultimately, a moratorium will only be fully effective if it is implemented, applied, and enforced domestically. Many of the countries in which the capacity exists to undertake germline genome editing already have in place rules relating to such matters, including prohibitions. We need to pay more attention to the adequacy, appropriateness, and effectiveness of these

---

[5] Further information on ARRIGE is available at: http://arrige.org
[6] Further details on the proposal are available at: https://www.globalca.org

rules, whether in the form of legislation, research guidelines, codes of practice, funding restrictions, or in other forms. This is not to deny that there are also other softer regulatory tools, including condemnation from the scientific community, adverse publicity, withdrawal of funding, and refusal to publish by academic journals. Nevertheless, we do need to be assured that appropriate penalties are in place to punish and deter conduct that is contrary to ethical norms and societal values.

More acceptable forms of conduct must be overseen and monitored to a level appropriate for the particular circumstances. We need to think much more deeply about whether the current regulatory regime for somatic cell genome editing is achieving this end. At its most basic, it must surely be designed to facilitate clinical applications of somatic cell genome editing that are socially, ethically, and clinically robust, and to ensure that the benefits are shared as widely and as equitably as possible. How we devise a regulatory environment that appropriately addresses questions associated with social and ethical robustness will require much deeper reflection and deliberation. The question of how distributive justice will be appropriately recognized and addressed is by far the most complex and most pressing issue for deliberation. One risk is that policy-makers might get caught up in the promise of genome editing and the alleged need to provide an incentive-based, light-touch regulatory architecture for bringing this technology into the clinic. We should not forget that other healthcare needs must also be supported out of a limited healthcare budget, nor that socially and ethnically disadvantaged members of society should be given the same opportunity to benefit from these advances in precision medicine. Genome editing should not be allowed to become yet another personalized service for the most advantaged members of society.

Acknowledgements

The research presented in this chapter is supported by a grant from the Australian Research Council, DP180101262.

15

The advent of automated medicine? The values and meanings of precision

Barbara Prainsack

Precision medicine as data-driven personalization

If we understand personalized medicine as a way of tailoring medicine to the individual characteristics of patients, then we need to acknowledge that important aspects of medicine have always been personalized. Specifics of patients' bodies, identities, practices, and circumstances of life have always influenced diagnosis and other aspects of medical thinking. What has changed, however, are the types of evidence and information that are used for this purpose (Prainsack, 2017; Prainsack, 2018). In the short time since the completion of the Human Genome Project (2000–2003) this change has been particularly rapid. In the early 2000s, individual differences in the form of genetic markers were seen as the key to stratification, and the matching of drugs to genetic markers of patients was the paradigmatic case of personalized medicine. Throughout the 2000s, this gene-focused approach gradually gave way to an understanding of personalization that included a much wider range of practices and information. This was the case, for one, because the understanding of the role of genetic factors in disease aetiology became more nuanced. It was made possible also by new opportunities to generate, integrate, and analyse different types of information. This new, data-rich, iteration of personalization, often subsumed under the label of 'precision medicine', focuses on the adjustment of prevention, diagnosis, and treatment to specific characteristics—genetic, behavioural, or otherwise—of patients (e.g. National Academy of Sciences, 2011; European Science Foundation, 2012).

The epistemic, technological, and practical shifts that gave rise to precision medicine are, in turn, linked to dominant values inscribed in larger economic and social transformations that contemporary societies are

Barbara Prainsack, *The advent of automated medicine? The values and meanings of precision* In: *Can precision medicine be personal; Can personalized medicine be precise?*. Edited by: Yechiel Michael Barilan, Margherita Brusa, and Aaron Ciechanover, Oxford University Press. © Oxford University Press 2022.
DOI: 10.1093/oso/9780198863465.003.0015

undergoing at the moment. In this chapter I argue that, against the backdrop of these larger societal developments, current visions of precision medicine are likely to lead to the polarization of healthcare. A combination of high-tech and high-touch medicine could be available for the select few who can afford to pay for the interpretation of their health data by humans, while the rest of the population may be expected to resort to automated medicine for most of their health concerns. In other words, only those who are wealthy enough to afford the time of human experts will be able to fully benefit from precision medicine.

My argument proceeds in the following steps: I start by outlining how current visions and practices of precision medicine bridge fields and perspectives that were largely separated in previous decades—a development that has been lauded as a new holism, also because of its reliance on system theory, and because of its connection between individual and population levels in clinical research and practice. This new holism, however, has reductionist elements, owing to its central orientation towards data. Not genes but data are the common currency of precision medicine in the twenty first century. This data-focus enables practitioners of precision medicine to see new things (e.g. formerly unknown correlations between health- and disease-relevant characteristics), but it also directs their gaze away from aspects that are not, or cannot, be captured in computable data. This can exacerbate existing social and economic divisions, and create new ones: divisions between wealthy patients on the one hand, whose ever more granular health information influences their own and other people's healthcare, and people who are missing from databases because they are not captured in patient registries and health insurance databases on the other. If we do not revisit and expand dominant understandings of precision medicine, we may end up with a new division within healthcare where expensive services will be offered to the rich that combine 'big data' approaches with 'big interpretation'—namely the time- and labour-intense practice of making sense of increasing volumes of data. The majority of the people would be offered a kind of automated medicine where the diagnosis and treatment of a wide spectrum of health problems would be done by machines. This can be avoided, however, if we expand the goals of precision medicine to include 'subjective' personal outcomes as well as social outcomes that are measured and rewarded alongside clinical outcomes. I argue that realization of just access to healthcare, and the improvement of social determinants, is a necessary precondition for greater personalization to benefit societies.

New connections in precision medicine

Proponents of precision medicine argue that it brings together aspects that were previously disconnected, and that it finds new connections that were unknown (e.g. Gillman and Hammond, 2016). These new connections include (i) systemic views that overcome disease-specific fragmentations in expertise and practice, (ii) linking individual- and population-level research, and (iii) integrating biomedical information with patient-reported 'subjective' input and information on social determinants.

Overcoming disease-specific fragmentation of expertise and practice

Less than two decades ago, in 2003, the historian and sociologist of medicine George Weisz observed that specialization in medicine was treated as 'a self-evident necessity of medical science whose existence requires little explanation' (Weisz, 2003, p. 538). While it still holds true today that medical specialties are considered an inevitable feature of healthcare provision, criticism of the fragmentation of practice and expertise, as well as calls for more systemic approaches to understanding and acting upon health and disease, are mounting (e.g. Stange, 2009; Bousquet et al., 2011).

Medical specialization had emerged throughout the nineteenth century in France and then travelled to other places in the Western world. According to Weisz (2003), an important driver of this process was the emerging scientific paradigm in medicine, which brought with it the idea that clinical researchers needed to specialize to be able to experiment rigorously and skilfully. Another factor was the increasing popularity of classification within the emerging administrative state, where the stratification of people into clearly delineated groups served as a means for the effective governance of populations. Specializations emerged partly around specific phenomena of the human life course, such as childbearing and sexuality (e.g. obstetrics and gynaecology, venereal diseases), and partly around a new disease taxonomy resulting from the development of new tools and technologies that were used to explore pathologies of, and within, specific organ systems (See Rosen, 1972 [1944], cf. Weisz, 2003).

Given that the first wave of medical specialization emerged at the intersection of a focus on life phenomena and new technological practices, then it is perhaps no surprise that the next wave of medical specialization would also be rooted in the emergence of new life phenomena and medical technologies. Precision medicine is, first and foremost, a phenomenon of the digital age, made possible by the increasing datafication of people's bodies and lives (Prainsack, 2017; Prainsack, 2018). New information about people's practices and preferences is often available in a digital format and can thus be used for data mining in the health context as well. This opens opportunities to create multi-omic datasets of patients, combining molecular and behavioural information. Using also non-obtrusive instruments and devices such as remote sensing and personal mobile devices, data can be collected in times of health and wellbeing and be compared to data collected in times of disease and illness, turning patients into their own 'control group'.

Next to facilitating digitization and datafication, new technological practices also play an important role in strengthening attention to the molecular level. The molecular characterization of physiological and pathological phenomena now complements—and in some contexts replaces—the organ- and disease-specific gaze that has characterized biomedicine in recent decades. New clinical trial designs focus on molecular characterizations of pathologies (such as tumours) rather than organ-specific phenomena (e.g. Menis et al., 2014; Simon, 2017; see also Schork, 2015). The US Food and Drug Administration approved their first site-agnostic oncological drug in 2017 (Lemery et al., 2017; Blumenthal and Pazdur, 2018). Immunotherapies in cancer (e.g. Couzin-Frenkel, 2013) further articulate an approach to health and disease that overcomes organ-specific fragmentation, and new institutes and experts on systems biology and systems medicine have emerged that pursue an integrative approach seeking 'to understand and predict the behavior or "emergent" properties of complex, multicomponent biological processes' (Strange, 2005, c968), and to translate these insights into preventive, diagnostic, and therapeutic applications (Clermont et al., 2009; see also Flores et al., 2013; Apweiler et al., 2018).

Connecting individual- and population-level research

As noted, contemporary visions of precision medicine mark a shift from (presumably) stable genetic and genomic information to the inclusion of wider ranges of malleable or dynamic data—including lifestyle and other behavioural data—for the purpose of personalizing research and healthcare. At the same time, they also denote a shifting understanding of difference

(Prainsack, 2015). In its extreme form, precision medicine assumes that people are different at so many levels that disease classifications need to be reconsidered. Traditional, disease taxonomies, such as the ICD, would then need replacement by data-rich characterizations of individuals at various stages of health and disease (National Academy of Sciences, 2011).

But if every person is her own standard, how do we know when a person's functioning ceases to be 'normal'? And how should we even define normal in a situation where there are no more shared diseases, but only individual manifestations of different configurations of characteristics and symptoms? Part of the answer to this is that individual health information, even if it is not directly compared to any population- or group-based standard, only ever makes sense in relation to others. Despite the rhetoric of personalization and individualization in medicine, routine clinical practice still operates with population averages. Physicians still refer to 'normal ranges' of various bio-markers in determining whether or not we need to be concerned, and many performance tests still compare us to groups that are defined according to old epidemiological categories. But even in cases where there is no fixed standard that determines what normal is in terms of a group average, seeing the results of others helps us to put our own results in perspective. We need to see the status of others to know what our own status means. This is true in research, where there is a continuous oscillation between individual-level data and population-level observations, but also in practice. In this context, it does not seem inevitable that personalization will give rise to a kind of atomistic individualism where patients are treated as unique individuals unconnected to others, and where patients also start to see themselves in this way (van Beers et al., 2018).

Connecting biomedical with personal and social factors

As noted earlier, one of the promises of precision medicine has been to overcome 'superficial' and fixed categories in the stratification of people (Prainsack, 2015). The old demographic and epidemiological labels used to 'stick' to people insofar as they related to characteristics that could not easily be changed, such as gender, age, class, ethnicity, place of birth. These stable labels are increasingly complemented by more fluid categories that attach themselves to molecular markers, behaviours, and other aspects that change along with people's bodies and practices. The greater capabilities for individual-level and dynamic characterization of people in molecular and behavioural terms also change approaches to studying disease causation. 'Big data' epistemologies include approaches that, instead of starting from functional pathways and checking whether a particular patient has a disease-causing characteristic, work the other way around: they start without a hypothesis and look for correlations between diseases and other characteristics. When such correlations are found, then the functional relationship between the disease and the corresponding characteristic is being scrutinized. In some contexts, the correlation alone will even suffice—especially when the resulting action would be a non-invasive measure (e.g. Mayer-Schönberger and Cukier, 2013).

This development helps clinical attention to go beyond somatic and bio-medical factors towards the inclusion of data and information on people's practices, preferences, and social and economic environments. The phenomenon of 'learning healthcare systems' is a case in point: it captures attempts to integrate high-scale data reuse and patient perspectives (e.g. Budrionis and Bellika, 2016). So far, whereas the latter have mostly been understood in terms of behavioural data—including data from mobile apps and social media—they increasingly include also 'subjective' factors. Patient-Reported Outcome and Experience measures (PROMs, PREMs) are but two examples. It is expected that patient-reported outcomes 'not only contribute to personalization of care, but also improve the accuracy of prediction models and enhance the decision-making quality of the clinicians' (Budrionis and Bellika, 2016, p. 89; see also Wysham et al., 2014). Rather than calling upon clinicians to consider patient-reported information only ad hoc, institutions, such as the Institute of Medicine within the US National Academy of Sciences (2012), have been calling for a systemic integration of these new measures in healthcare. Learning healthcare systems should include intelligent automation, comparative effectiveness research,[1] positive deviance (i.e. bottom-up, community-based learning from cases of exceptional performance), surveillance, predictive modelling, and clinical decision support (Learning Healthcare Project, 2019).

---

[1] Comparative effectiveness research compares groups of patients to explore what factors correlate with better outcomes. As such, this approach is not new, but the scale upon which it is funded is novel (e.g. Garber and Tunis, 2009; Khoury et al., 2012), as are the ranges of treatments and interventions that it looks at. Moreover, comparative effectiveness research in the USA at least is now including patients' and other stakeholders' perspectives upstream, instead of being driven exclusively by healthcare providers (see Selby et al., 2015; Gargon et al., 2018). As a result, studies are now under way that include social determinants in comparative effectiveness research (e.g. McClintock and Bogner, 2017).

This growing attention to data and information outside of the realm of bio-medicine is partly due to the push of technological advances that create data as a surplus of automation and digitization—meaning that processes created to take certain things from the analogue to the digital world, or from humans to machines, are generating additional layers of data as a byproduct of their function. An example is medical imaging, where semi-automated quantifi-cation by imaging software is generating data that used not to be available in the days when images were analysed by human 'eyeballing' only. Similarly, a mobile application on a patient's smartphone or tablet where she is supposed to track her moods, or the side-effects of a medication, is capturing all kinds of collateral data—such as GPS information, data on when she is awake and active, etc. These data could be mined in its own right.

Besides the push of technology, there is also the pull of economic factors. In some cases, it could "pay" to consider more systematically 'subjective' per-sonal factors and also social and economic determinants within healthcare. In the context of the rise of chronic diseases and co-morbidities, as well as residual factors that increase healthcare costs, there is increasing pressure to use (and reimburse) *whatever works*—even if the effective measure in ques-tion is not a biomedical intervention in the traditional sense. If massage ther-apies, hypnosis, or nudging instruments bring the desired outcome, there is an increasing demand to make them available to patients, especially when they are cheaper than the alternatives.

New divisions: the enlightened reductionism of precision medicine

While precision medicine helps to bridge disciplinary and epistemic divides, it also contributes to the creation of new gaps. Most of these emerging rifts are—directly or indirectly—connected to a new reductionism entailed by the data-centrism of precision medicine. It is reductionist in the sense that it marginalizes, or even excludes, evidence that is not readily available as data—such as verbally reported experience (see, e.g., Chisholm et al., 2020). At a time when data is the common currency of the medical gaze, it rearticulates these in terms of data. In in the context of integrating patient perspectives into clinical decision-making (for example Budrionis and Bellika, 2016), ra-ther than looking for ways to include patient narratives and other types of 'thick descriptions' of what patients feel, hope, and care about, patient per-spectives are limited to readily available behavioural data from mobile apps, social media, and telemedicine applications. Similarly, when research oscillates between individual- and population-level information, digital data are often the denominator that makes them commensurable. Finally, also in the bridging of disease-specific expertise and practice, data form the central lens through which human systems are viewed: the body is first taken apart into data points and then put together again. This data-centrism is not unique to precision medicine, but it is a characteristic feature, predicating the type of holism that it fosters. This new holism, as Vogt and colleagues argue, is no longer primarily oriented towards humanistic ideas, but it represents patients in terms of data and seeks to render them predictable:

As Henrik Vogt and colleagues argued, each person's whole life process is defined in biomedical, technoscientific terms as quantifiable and controllable and underlain a regime of medical control that is holistic in that it is all-encompassing. It is directed at all levels of functioning, from the molecular to the social, continual throughout life and aimed at managing the whole continuum from cure of disease to optimization of health (Vogt et al., 2016, p. 308; see also Vogt, in this volume).

Digital data is the epistemic and practical unit that enables connections between different territories of research and practice; the digital data point has become the universal unit of analysis that enables the integration of different domains of our life, and different aspects of our bodies. It is the unit for which learning healthcare systems create data architectures and highways and other technological and social infrastructures, and that attracts public and private funding. Circumscribing Adam Hedgecoe (2001), who used the term 'enlightened geneticization' to describe how genetic causes are given priority over other factors in explaining narratives of schizophrenia, what we observe within current visions and practices of precision medicine is a kind of 'enlightened reductionism' that takes a systemic and holistic perspective through the lens of 'data'. It sees new things, but it also blinds itself to key aspects that are not already datafied.

The blind spots of precision medicine

In the era of data-rich medicine, concerns about bias have become prominent in expert and public debates alike. Bias can stem from a wide range of social and technological practices; we can roughly separate them into statistical and social biases (Parikh et al., 2019). For example, the training datasets for an algorithm that supports diagnosis, these could underrepresent certain groups of people for whom the resulting diagnostic algorithm then does not work. Addressing this problem is complicated by the fact that there is often no reliable, properly labelled information available that can help to discern and detect bias. Frequently there is also conflict between two different goals: the use and retention of information on protected characteristics such as gender and race on the one hand, which are necessary to assess whether a software application is biased against certain groups, and the attempt to protect people and groups that mandates the removal of such labels on the other—because these labels have led to discrimination (e.g. Veale and Binns, 2017). Moreover, biases in social and healthcare practices and institutions have contributed to long-standing discrimination against women and minority groups in medicine and healthcare. Data-rich personalization may not remove biases, configure them differently. It may even enhance bias. For example, because 'black' women are less likely to be diagnosed with depression than 'white', even if they show symptoms of depression (e.g. Carrington, 2006), this bias might affect the training data-sets for algorithms helping to detect depression. This might be the case even if 'black' women were not underrepresented in the training data. The more data-rich characterizations of individuals diffuse into healthcare, the more such bias is at risk of affecting personalized care. And when it is believed that machines are less prone to bias than humans are, the awareness of possible bias in algorithmic systems is lower (see also Pot et al., 2019).

A new class division in healthcare? Boutique versus automated medicine

The unintended discrimination that can emerge from data-driven personalization is not the only way in which precision medicine could foster new divisions. There is a risk that one of the most expensive parts of precision medicine, namely data interpretation—including the human contact that is required to establish what 'makes sense' for a patient and how—will be available only to the selected few, while the masses will have to make do with the help of machines for all but the most serious and acute of health issues. In other words, precision medicine would be divided between automated care and boutique practice.

The groundwork for this process, in some countries, has already been laid. In the USA, in the early 2000s, boutique (or concierge) medicine (Reinhardt, 2002; Clark et al., 2011) emerged as a model within which physicians offer bespoke services to a very limited number of patients who pay a fixed annual fee regardless of the services rendered (retainer). By the end of 2019, about 12,000 US physicians were reported to have opted out of other arrangements and into boutique medicine, charging their patients between US$125 and 200 per month (Daily, 2019). Besides being available for their patients around the clock (and often also by phone or text message) and offering same-day appointments, boutique doctors also facilitate high-tech testing and analysis including DNA sequencing and cutting-edge imaging technologies. They also review results together with their patients. Many healthcare professionals working within the traditional fee-for-service model do not have the time to sit down with patients to interpret data together, neither do they have the time to train and educate themselves to keep abreast with the latest technological developments. The difficulties involved in interpreting data, and the time, effort, and skill needed to do so, is one of the untold stories of precision medicine (see also Reardon, Chapter 16).

The more money invested in digital infrastructures, including patient monitoring and preventive testing, the greater the need will be for healthcare professionals who can help patients make sense of the resulting data and information. But the human power and time to do so is expensive. The boutique model could seem to be the only way in which doctors can afford the time and effort necessary to do the 'data work' (Bossen et al., 2019)—especially in a context where this task requires much more than the reading of lab results and determining whether a specific patient is within a normal range of healthy values. In precision medicine, because many types of data are novel, not (yet) standardized, or ambiguous, 'normal ranges' have not yet been established; the problem of 'variants of unknown significance' in genomics is a case in point (e.g. Ackerman 2015; Eccles et al., 2015). Practices such as whole-body imaging pose similar problems as there is not yet sufficient information from a large enough number of other patients to understand what it means (e.g. Agriantonis et al., 2008; Morin et al., 2009).

Some people put their hope in computers that mine large datasets to discern patterns. There are expectations that machines will help with risk prediction, diagnosis, choice of intervention and follow-up (Obermeyer and Emanuel, 2016; Shameer et al., 2018; Topol 2019). None of the proponents of machine learning in clinical medicine expect that machines fully replace humans; just as there are things that machines do better than humans, there are things that humans do better than machines. If high-quality healthcare is the goal, then human minds will be needed to work alongside machines—e.g. to explore whether a specific factor found to correlate with an outcome could be a potential cause.

Here is where the larger context of healthcare comes in. In the context of rising cost pressures and the need for solutions that work 'well enough', despite knowing that machine solutions for data interpretation are fallible, they will work 'well enough' for most patients. And at the population level, "well enough" may be seen as the most cost-effective approach. The "outliers" for whom these automated approaches do not find good solutions for, might be left without adequate care, or worse, harmed by machine errors. Already disadvantaged minorities are also at risk to be among those left behind.

Table 15.1 illustrates the scenario of an increasing division between boutique medicine, which would give people who can afford the monthly fees access to a wide range of high-tech tests and high-touch services such as annual check-ups, health coaching, and other perks. For the 99% of the population who cannot afford boutique options, automated medicine would become the default: when patients have problems or questions, their first point of contact would be a website or telephone line where an algorithm and chat bot asks questions and helps diagnose their problem. If the problem were considered serious and acute enough, they would be routed to a first responder unit staffed with people who decide what intervention the patient needs. If the problem seemed acute and serious enough, the patient would be assigned to a clinic where healthcare professionals would see her in person; if the problem did not seem acute enough to merit human intervention, the patient would be referred to other online services that carry out testing or monitoring. Medication needed would be delivered to the patient by mail, and monitoring would be done via apps that patients are told to install on their personal devices. If patients required someone to talk to them, a digital

Table 15.1 A new type of class-based medicine

	Cost	Level of technology involvement	Cost of technologies involved	Involvement of humans	Type of personalization
Boutique medicine	High	High	High	High	Digital and narrative, holistic
Automated medicine	Low	High	Low	Low	Digital, reductionist

'therapist' would be assigned to them who would be so well trained that it would be indistinguishable from a human in direct interaction.

Conclusion

Against the backdrop of discourse of rising cost pressures and technological development that renders it impossible to provide the highest quality of healthcare to everyone, data-driven 'precision' medicine is likely to develop in the direction of a new two-tier medicine. This scenario can be avoided, however, if we expand the prevailing paradigm regarding precision. One of the consequences of precision medicine's data-focus are new correlations established by data-mining approaches. With this comes a greater pressure to put into practice 'whatever works', even if the new correlation—between a lifestyle factor and a disease, or a specific intervention and an outcome—is not well understood, or does not square well with established theoretical knowledge. In turn, this development, in turn, means that more attention is paid towards social or personal factors that have bearing on clinically relevant parameters and outcomes. The improvement of access to safe and affordable housing, food, transportation, education, and healthcare—as suggested, for example, within the Universal Basic Services model promoted by the Institute for Global Prosperity (2017) at University College London—would go a long way in removing or mitigating factors that make people ill, including social isolation. An understanding of precision medicine that includes aspects in the personal, social, and economic environment of the person as factors that bear on health outcomes, and also pays specific attention also to the social determinants of health, may facilitate sustainable and fruitful investments in healthcare, as well as prevent the widening of social inequalities. To reach this goal, it will be necessary to look beyond person's 'behaviour'—in terms of the food she eats and other lifestyle 'choices' of the person—as information to be considered in prevention, diagnosis, and treatment decisions, and consider also the social and economic determinants that shape these patients' 'behaviour'. We need a precision medicine that measures its success and its effectiveness not only in terms of individual clinical outcomes, but also in terms of health, wellbeing, and equity at the level of populations.

Acknowledgements

I am grateful to Yechiel Michael Barilan, Margherita Brusa, and Aaron Chiechanover for organizing the meeting that led to the writing of this chapter, and for helpful comments on the manuscript. I would also like to thank Carrie Friese, Hanna Kienzler, Yechiel Michael Barilan, and Margherita Brusa for their insightful advice on how to improve an earlier version of this chapter. The usual disclaimer applies.

16

Thoughtful genomics

Jenny Reardon

The manifestation of the wind of thought is not knowledge; it is the ability to tell right from wrong . . . And this, at the rare moments when the stakes are on the table, may indeed prevent catastrophes. (Arendt, 1978, p. 193)

Introduction

In the opening pages of *The Human Condition*, the public intellectual and political theorist Hannah Arendt made what she described as a 'very simple' proposal: 'It is nothing more than to think what we are doing' (Arendt, 1998, p. 5). Here, inspired by Arendt, I make another simple proposal: to create a thoughtful genomics, one with the capacity to think and speak about that which it does. I do so in light of what I argue is the most significant ethical problem posed by genomics: the problem of thought.

While there is not space here to adequately explore what I mean by thought, sufficient to say that I am Arendtian in my belief that thinking poses the problem of meaning. Arendt articulates this position in her Gifford Lectures published at the end of her life under the title *The Life of the Mind*.

Even the relentlessness of modern science's Progress, which constantly corrects itself by discarding the answers and reformulating the questions, does not contradict science's basic goal—to see and to know the world as it is given to the senses—and its concept of truth is derived from the common-sense experience of irrefutable evidence, which dispels error and illusion. But the questions raised by thinking and which it is in reason's very nature to raise—questions of meaning—are all unanswerable by common sense and the refinement of it we call science. (Arendt, 1978, pp. 58–9)

Jenny Reardon, *Thoughtful genomics* In: *Can precision medicine be personal; Can personalized medicine be precise?*. Edited by: Yechiel Michael Barilan, Margherita Brusa, and Aaron Ciechanover, Oxford University Press. © Oxford University Press 2022. DOI: 10.1093/oso/9780198863465.003.0016

She concludes:

> [I]f [we] were ever to lose the appetite for meaning we call thinking and cease to ask unanswerable questions, [we] would lose not only the ability to produce those thought-things that we call works of art but also the capacity to ask all the answerable questions upon which every civilization is founded. (ibid., p. 62)

Arendt came of age during the rise of Hitler and National Socialism, and experienced first-hand the terror wrought by mobilizing calculative modes of knowing to justify the regime's policies of eugenics, and its extermination of millions of lives. She worried specifically about the takeover over of the natural sciences by informatics and mathematics. Mathematical truths, Arendt argued, did not 'lend themselves to normal speech and thought', and thus threatened to erode capacities to 'think and speak about the things which nevertheless we are able to do' (Arendt, 1998, pp. 3–4). The idea that mathematics should serve as the 'paradigm for all thought,' she recognized, was as old as Pythagoras (Arendt, 1978, p. 59). Yet she worried that knowing based in abstract calculations might lead human beings down the destructive path of adhering to logics that were disconnected from 'facts and experiences'—a *logicality* that fostered totalitarianism (Arendt, 1994, p. 358; Fins and Reardon, 2019).

Arendt distinguished this kind of *knowing* from what she defined as *thinking*, a distinction she grounded in the Kantian delineation between *Verstand* (the ability to know, or intellect) and *Vernunft* (speculative thought and reason). The former sought irrefutable truths. The latter fostered the discernment of the meaning—the value, significance and good—of efforts to know (Arendt, 1994, pp. 57–58). A collective political life, Arendt argued, depended on this latter capacity to reflect together about the value and significance (i.e., the good) of efforts to know. It required, in the parlance of chemist and science studies scholar Isabelle Stengers, *public intelligence*—an intelligence born out of 'a process of hesitation, concentration and attentive scrutiny' to the reality that living together across difference requires answering questions for which there is no single right answer, but rather a series of difficult choices (Stengers, 2018, 4). Collective capacities to not destroy, but rather to love the world—to want it to *be*—she believed, required defending spaces not just for definitive truths, but for this speculative thought. Such thought fostered a love of the world that sought not material possessions possession (*cupiditas*), but rather understanding (*caritas*) (Paine, 2013).

The task of cultivating a thoughtful public intelligence with the capacity to make hard decisions across human differences today presents the field of human genomics with one of its most important challenges. Here I briefly describe why and how this became the case, and how a thoughtful genomics might be forged.

Genomics' problem of meaning: historical antecedents

Over the last three decades, molecular biologists, computational scientists, and mathematicians have built machines and data platforms with the power to sequence and store an exponentially growing number of human genomes. Their success rightly has been celebrated for revolutionizing the life sciences. However, as genome scientists sharpened their technical acumen and abilities to create genomic data, they set to one side hard decisions about what that data could or should mean. As I have argued elsewhere, there was a decisive moment in the history of genomics when leaders of the Human Genome Project (HGP) decided that the goal of the project was only to accurately sequence the nucleotides of one human genome. If in the nucleus of a human cell the beginning of the AADACL3 gene on chromosome 1 reads CAT, then the goal of the HGP was to ensure that it read CAT—not CAG—when it was digitally transmitted to a computer screen. The task was not to decipher what the CAT meant, but rather to precisely encode these nucleotides in digital form. The task of meaning-making, they argued, should be left to future generations (Reardon, 2017).

This decision to encode—not decode—the human genome fundamentally shaped the field of genomics. Following the completion of the HGP, leaders of genomics continued to sidestep thinking about how and whether genomic data should be embedded into health systems, criminal justice databases, and insurance formulae, and how they should or should not design sampling strategies and data platforms to facilitate or hinder these different ends. Instead, they continued to focus energy on creating ever more powerful sequencing machines, cloud platforms, and algorithms that could create, store, and compare genomic data to other stores of data. A genomic apparatus thus formed which was founded upon the belief that a calculative mode of knowing—'big data' informatics—could produce knowledge about human bodies and bypass the messy interpretive work of human decision-making. Indeed, advocates celebrated the elimination of social and cultural bias through the replacement of human beings and their faulty powers of judgement with big data and powerful machines (Reardon, 2017, ch. 5).

While this approach helped the field to speed ahead, it also hid from view the human choices that attend the construction of big data sciences like genomics (see Barilan, Chapter 7). The kinds and amounts of data that one *could* collect are endless. How then do 'we' make a *de-cis-ion* about what categories should orient limited attention and capacities for care? Along with Arendt, I hyphenate *de-cis-ion* to call attention to *cis* which lies in the middle of the word. *Cis* comes from the Latin *caedere*, which means to cut or to kill. As we decide what data we will collect, we are also deciding what data we will not collect. Every observation, as Arendt explained, is an act of perception *and* misperception (Arendt, 1978, p. 54). It cuts some things from our view, while bringing other things into focus. To see is to decide. To decide is to judge what 'we' *ought* and *ought not* to observe and know—and thus, what we ought to *do* or not do. At stake is not just the truth, but decisions about whose categories—and thus whose values—will order the data infrastructures that guide our lives, and make 'us' up.

Most notably, since genomics' inception, the field has been dogged by questions about whether it should collect data about 'race' and 'ethnicity', and, if so, how those categories should be defined (Lee et al., 2001; Reardon, 2005). While genome scientists have repeatedly promised that they would transcend group categories—and provide information directly about an individual person—as Barilan explains in this volume, personalized medicine bases its analysis on comparisons. One is never just an individual, but an individual understood in terms of other individuals grouped into risk categories. Despite genome scientists' ongoing efforts to create 'precise' categories that are devoid of 'social' content, in practice many widely used genomic databases use ethnic and racial categories, such as 'European' and 'African American' (Reardon, 2007).

The choice to use these categories affords individuals and states new possibilities for conducting racial profiling and asserting racist beliefs. The Chinese government, for example, uses genomic data to target ethnic minorities for surveillance (Wee, 2019). Law enforcement agencies increasingly use genomics in forensics work to make inferences about the race and ethnicity of samples left at crime scenes (M'charek et al., 2020). Leaders of the alt-right invoke results from genome testing services like 23andMe to celebrate their racial purity (Panofsky and Donovan, 2019).

Yet despite these realities, genome scientists largely have been reticent to discuss the relationship between genomics and issues of race and racism.

New York Times reporter Amy Harmon, for example, found very few genome scientists who were willing to go on the record to speak to her about her 2018 article about white nationalists' appropriation of genomic data. As Harmon explained, some believed their duty was to produce the data, not to weigh in on its meaning. Others thought that those who made such claims were fringe members of society, not worthy of comment. Still others thought that in fact there might be correlations between genomics and race (Harmon, 2018). For these reasons, of the many with whom Harmon spoke, only a few agreed to speak publicly.

Arendt was deeply concerned about this kind of collective reticence. She argued that 'thought-words'—or speech—enabled collective life, and staved off devolution into brute rule by totalitarian demagogues. It made 'human to-getherness' possible. Speech, in this sense, was a form of articulation (Arendt, 1998, p. 180). It joined. For Arendt, the stakes of thoughtful speech were high. She famously attributed the slaughter of millions in the concentration camps to a logicality that arose from thoughtlessness (Arendt, 2006).

While we might sincerely hope such a massive atrocity is no longer possible, seventy years later, 'group think' and demagoguery remain central concerns. Indeed, their potential power has expanded. Today big-data platforms and algorithms augment radio waves, deepening the potential reach, depth, and automation of mass-scale messaging (Noble, 2018). Further, whereas seventy years ago many geneticists were willing to think critically and speak out about their field's entanglement with ideological forces, today most genome scientists cling to the belief that they do the most good by honing their technical expertise to produce more data and powerful algorithms (Brattain, 2007).

Yet calculative knowledge alone remains as impotent as ever if not harnessed to collective critical thought. If there is no deep thinking that goes along with the 'deep medicine' (Topol, 2019) promised by genomics, then the field will foster an intellectual purgatory in which thought is suspended, and brute rule by demagoguery may replace deliberative reflection and decision-making.

Ethics' problem of meaning

For decades leaders in genomics turned to ethics to avert this end. In 1993, when the US Congress created the National Center for Human Genome Research (NCHGR), it mandated that not less than 5% of funds be devoted to 'the ethical, legal and social implications of genomics'. Genome scientists did not resist the devotion of such significant funding to ethics, but instead embraced it. They understood that their field raised significant societal and ethical issues (e.g. privacy, social discrimination, commercialization). As first director of the HGP James Watson put it, 'the cat was out of the bag' (Juengst, 1996, p. 66). Thus, the best thing to do was to address the issues head on.

While many lauded the decision, others worried that the newly created Ethical, Legal and Social Implications (ELSI) Program would create nothing more than 'a screen of ethical smoke' (Juengst, 1996, p. 67). Whatever one's view about the sincerity of the effort, it was clear from the start was that the ELSI Program would not address questions about the value and meaning of human genomics. As its first director, bioethicist Eric Juengst, explained in his reflections published in 1996, 'the enterprise of genome research and the knowledge generated by it' were to be treated as 'unalloyed prima facie goods' (Juengst, 1996, p. 68). The job of the ELSI Program was to create the policy tools that would ensure that these goods reached the public in the most efficacious and equitable manner. If the science was revolutionary, then the ethics was to be accommodating.

As a result, rather than raising questions about the role genomics played in reconstituting medicine and social order, the work of the ELSI Program focused on adapting existing ethical practices (e.g. informed consent) and legal instruments (e.g. patent law) to accommodate the new domain of genomics. When questions did arise about genomics' role in ordering and interpreting human differences—specifically, in making claims about race—leaders of the HGP argued that those issues were not germane. For example, when *The Bell Curve* was published, prominent sociologist and member of the ELSI Working Group, Troy Duster, requested that the American Society of Human Genetics evaluate the books' claims about race, genetics, and IQ. However, Francis Collins, then head of the HGP, resisted (Duster, 1995). Because the book addressed behavioural genetics, Collins argued, it had no relevance to the work of the HGP. The HGP, its leaders maintained, sought to create the tools and instruments to sequence one human genome. It did not address questions about the order of the human species, or about meaningful human traits, such as IQ. Those questions, they believed, should be addressed only after the HGP had been completed, the tools created, and the data produced. In the meantime, they funded ELSI scholars to address questions they considered immediately pertinent—for example, questions about who could own and control genomic tools and how. According to the leaders of the HGP, Juengst explained, the goals of ELSI was to create 'the tools and infrastructure' to address these policy questions (Juengst, 1996, p. 86). Within the HGP, ethics was understood to be an instrument designed to achieve goals whose value and meaning were beyond critical reflection. It was not a framework for reflecting about the desirable, the good and the meaningful.

At the time, there were those who questioned this tool-maker approach to ethics (and science). Among them were leading theologians within the Catholic Church. Rocked by strife over the proper uses of one of the very first forms of the biotechnology—the birth control pill—questions about the meaning and value of innovations in the life sciences had become topics of fierce debate within the Church. In 1968, Cardinal O'Boyle of Washington DC disciplined 41 priests for opposing Pope John Paul VI's teachings on the pill presented in the papal encyclical, *Humanae Vitae* (Cajka 2021). That same year, Jesuit and physicist Robert Brungs and biochemist John Matschiner co-founded the Institute for Theological Encounter with Science and Technology (ITEST) to provide space for these debates, and for considering the meaning of advances in science and technology for the Catholic faith. Located at St Louis University (where both Brungs and Matschiner served as professors), ITEST was located down the street from one of the largest human genome sequencing labs in the world, Bob Waterston's lab at Washington University. Taking advantage of this close relationship, in October of 1996 ITEST hosted a workshop entitled *Patenting of Biological Entities* (Institute for Theological Encounter with Science and Technology, 1996).

At the time, patenting had moved to the heart of ethical debates about the HGP. In 1995, Myriad Genetics attempted to patent the breast cancer genetic variants, BRCA1 and BRCA2. Jeremy Rikfin, technology critic and then President of the Foundation for Economic Trends, mobilized a response, calling for a ban on gene patenting (Hall, 1996). Hundreds of religious leaders joined in support. However, ITEST's Robert Brungs was not among them. Indeed, in an essay he wrote in the lead-up to the Fall 1996 meeting Brungs questioned whether patenting should be the main concern. The focus on rightness or wrongness of a particular technique such as patenting, he argued, obfuscated important questions of meaning. As he explained:

> Are we certain that lines should be drawn where we would at present draw them? The problem we face is not ethical; it is ontological, if I may use that word. Our issues and challenges are at the level of meaning. Ethical discourse is necessary, but it is in no way sufficient. (Brungs, 1995, p. 14)

He went on to cite a portion of the address that Harry Boardman, then Secretary-General of the Council for Biology in Human Affairs at the Salk Institute, delivered to the 1976 meeting of the American Association for the Advancement of Science (AAAS). Under the title, 'Some Reflections on Science and Society: A Terrain of Mostly Clichés and Nonsense, Relieved by the Sanity of Whitehead,' Boardman argued:

> [F]ar too pervasively, these endless biomedical-science-value discussions manifest a deplorable blindness which seems to proceed from a hypnotic fascination with appliances and appliance-makers. . . . The central concern is not science or scientists, but with the whole of knowledge—its benefits, the price it extracts, and its special province; that of ideas. (Ibid., p. 15)

The Salk Institute—founded in 1960 by the inventor of the polio vaccine, Jonas Salk—sought to bridge the sciences and the humanities. Described as an 'experiment in the sociology of science', it represented an effort to create the kind of thoughtful engagement across differences that Arendt imagined (Wade, 1972). In the 1970s, science studies pioneers Bruno Latour and Steve Woolgar spent time there and studied the laboratory of one of the Institute's main founders, the neuroscientist Roger Guillemin. Their results appeared in *Laboratory Life: The Construction of Scientific Facts*, a ground breaking work that provided an entirely new way of interpreting the meaning of the calculative instruments of science, revealing them to be forms of power (Latour and Woolgar, 1979). In his address to the AAAS, Boardman is advocating for a less myopic view of science that engages these more fundamental ontological stakes of science. 'In other words,' Brungs explains, 'Boardman is calling for less attention to techniques (and to issues like patenting) and more concern about the meaning of these new powers' (Brungs, 1995, p. 15).

The discussions that were to follow at the Fall 1996 ITEST meeting brought into focus the importance of attending to these issues of meaning. It also illustrated the tendency of genome scientists to skirt around them. At that meeting, Bob Waterston described how commercial companies—from biotech and pharma—were positioning themselves in the domain of genomics. At the time, Waterston led one of the two main laboratories that was producing the final sequence of the human genome. John Sulston led the other at the Sanger Institute in the UK. Both Waterston and Sulston were deeply committed to keeping the sequence in the public domain (Sulston and Ferry, 2002). Yet at the same time Merck & Co., Inc. funded Waterston's work. At the meeting, Waterston reflected on this apparent contradiction.

Merck came to us because they had made the corporate decision that they wanted to sponsor a project that would put on the Web human DNA sequence of the same kind that Incyte and Human Genome Sciences were producing. They wanted to sponsor projects with equivalent information out in the public domain. The reasons they wanted to do this, they never told us. I can only speculate and I would rather not. I think it is basically that they wanted to level the playing field. (Institute for Theological Encounters with Science and Technology, 1996, p. 155)

Waterston's desire to not probe into the reasons for Merck's support is illuminating. At the time, HGP leaders were building the ethical value of their endeavour largely on the grounds that they sought to keep Craig Venter and the company he helped to found—Human Genome Sciences—from locking the human genome up in patents. Merck's support of Waterston blurred the bright ethical line that the National Institutes of Health sought to draw between what they viewed as Venter's self-serving, private effort to sequence the human genome and their own public effort to sequence the human genome for the common good. It complicated the image of HGP scientists as 'white knights' who sought to protect the people from corporate interests (Reardon, 2017, p. 35).

Line drawing—between private and public—as Brungs warned, failed to hold. Instead, the very effort to draw a line raised critical questions about the constitution of public and private science, and what it meant to engage in the former and not the latter. What did it *mean* for Merck, a private large pharmaceutical company, to support the public effort? As theorist of biocapital Kaushik Sunder Rajan argues, Merck sought to ensure that biotech companies like Human Genome Sciences and Celera did not capture the rights to control genetic data (Sunder Rajan, 2006). Such a capture would have impeded the ability of Merck to freely use the data to innovate new medicines.

Who, though, would ensure that the public could access the medicines Merck created? What practical value would genomic data have if people could not afford the resulting medicines? Leaders of the HGP skirted around these questions. Instead, they argued that placing genomics in 'the public domain' was an unequivocal good.

The consequences of failures to address these questions of meaning today are becoming evident. As the last few years have made clear, some of the medicines arising from genomics are the most expensive ever produced. Spark, for example, released the gene therapy Luxturna to treat retinal degeneration in October of 2018 at a cost of $425,000 per eye, or $850,000 per person (Scutti, 2018). In 2019, the US Food and Drug Administration approved Novartis' release of Zolgensma, a gene therapy for a rare childhood neuromuscular disorder, at a price tag of $2.1 million (Stein, 2019). Later that year, the *New York Times* reported on several more multi-million dollar drugs for rare genetic disorders that threatened to bankrupt healthcare systems (Thomas and Abelson, 2019).

Rather than a beacon for public-minded science that benefits all humanity, increasingly genomics and its medical offspring, personalized medicine, are cited as causes of growing inequalities in medical care. In San Francisco, birthplace of the biotech revolution, residents recently asked leaders of University of California, San Francisco (UCSF) how they could invest so much money in the new Mission Bay campus—a campus devoted to the launch of precision medicine—while defunding community health clinics (Reardon 2017, p. 171). In 2016, Sandy Weill, American banker and the architect of the dismantling of the Glass–Steagall Act, gave 185 million dollars to UCSF to build a new neurosciences institute that would leverage genomic technologies (Farley, 2016; Norris, 2020). At the same time, UCSF threatened to close New Generations, the last full-service reproductive health clinic serving poor Black and Latino youth in San Francisco. As a petition posted to the Color of Change.org put it: 'The message the city of San Francisco and the University of California San Francisco are sending is blatant: money matters, not Black and Brown lives' (Barros, 2016).

Today such concerns about the entanglement of genomics and personalized medicine in logics of racial exclusion are on the rise (Saini, 2019). Leaders of genomics respond—as they did in the past—by arguing that the concerns are misplaced. Specifically, they contend that genomics will lead to the transcendence of group categories, and the creation of *personalized* and *precise* medical care that will help end racism by attending to an individual's precise genomic make-up. Yet, as Margherita Brusa and Donna Dickenson (Chapter 8) and Michael Barilan (Chapter 7) argue in this volume, such claims are based on idealizations detached from reality. In practice, human group categories will remain central to genomics' production of risk stratification. The ongoing concern will not be *whether* human group categories, but *which* human group categories.

A belief that technological innovation will save us from such concerns undermines capacities to respond to fundamental questions: What kind of world to do we want to live in, and what categories should we use to order it? What ought to be a thing around which we gather? *Thing* as the French theorist of technoscience, Bruno Latour, reminded us, derives from the Old English word 'ding', meaning a meeting or assembly (Latour and Weibel, 2005). A thing, or a gathering, arises around a matter of collective concern. Should a 'genome' be such a thing? Should 'population'? Should 'race'? By placing the focus on creating sequencing machines—the appliances, to use Boardman's language—the architects of genomics failed to build the infra-structure needed to address these critical questions. As a result, the field risks fostering an unthinking logicality that could inadvertently foster brute rule by capitalist and racial logics.

How can this be averted? How can 'we' think?

Cultivating genomic thought

Some insights might be gained by asking Arendt's question: *Where are we when we think?* (Arendt, 1978, p. 197).

Long ago people stormed hallowed monastic halls, demanding that the critical activity of thinking not be the special preserve of those who lived in monasteries in retreat from society, but rather those who were a part of it. The results were revolutionary. The Church would no longer control future Galileos. Interpretation—meaning—would arise from difference sites and sources, including scientists. The Catholic Church's own Pontifical Academy of Sciences, whose headquarters provided the space for the development of the thoughts collected in this volume, provided one important new venue.

With these new sites and sources of thought came new *response-abil-ities*. Inspired by Donna Haraway and Karen Barad, by this I mean that they brought obligations to cultivate capacities to respond to the world created from the new powers to think (Haraway, 2016, p. 71; Barad, 2007). The modern history of science is replete with examples of scientists struggling to decide what form this response-ability should take. Perhaps the most famous case is that of J. Robert Oppenheimer and Hans A. Bethe's struggles over their roles in creating the atomic bomb (Schweber, 2007), the creation of which forever unsettled the notion that the laboratory was a pure space that existed apart from society. We are, as Arendt called us to recognize, in place when we think. Thinking calls us to move ourselves 'from what is present and close at hand' in order to sense that place, and the difference—the meaning—it makes (Arendt, 1978, p. 199).

A thoughtful genomics would recognize that this new domain of technoscience is not a panacea capable of revolutionizing medicine in all cases and places. Instead its tools and diagnostics operate and are effective in *particular* cases and places. As physician Joseph Fins (Chapter 10) and clinical geneticist Christopher Austin (Chapter 9) explain, rare cancers and neurological diseases are areas where molecular diagnostics and biomarkers have transformed diagnosis. The question is how far do these cases extend? Advocates of personalized medicine proclaim that their field will revolutionize *all* of medicine. This totalizing claim undermines collective abilities to understand the situatedness of genomic practice, and how social action and critical reflection can make the genomic approaches to medicine more or less possible in different contexts. It hides from view decisions made that facilitate some modes of knowing and caring, and not others. Data may be endless. However, biological knowledge requires endings (Reardon, 2019). Judgements must be made about what data to collect and not collect and for what ends. The question is not whether to judge, but how. As Arendt reminds us, there is perhaps no question more fundamental to political life than this one: On what grounds will we judge, and who are 'we'? (Arendt, 1998, pp. 217–18)

Too often the 'we' of personalized medicine is made up of those who have the means to pay, and who trust the medical system. Those who have ongoing concerns about medical and scientific racism are unlikely to participate in personalized medicine initiatives like the USA's *All of Us* program (Lee et al., 2019). And even if *All of Us* successfully collected genomic data from enough of us to adequately represent the genetic variability of the nation, significant challenges remain (Hayden, 2020). The ongoing high costs of drugs, and the lack of market incentive for investment banks to back expensive personalized drugs with small markets, mean that only some today have access to personalized treatments (Kim, 2018).

Finally, even if all people could receive treatments, should that be the goal? This was the question raised by UCSF's effort to close the New Generations health clinic in the midst of its massive investments in personalized medicine. Barbara Prainsack (Chapter 15) urges that we should not forget that some of the most effective solutions for people's health problems are thus the simplest, low-tech ones: fight social isolation.

A thoughtful genomics would re-member (Arendt, 1998, p. 315). Arendt hyphenated 're-member' to remind us that as we remember, we decide who is and who is not a member of our polity. Precision medicine may forget too many people's health problems—leaving them outside a polity where healthcare is personalized—because its new buildings and spaces are not built for the vast majority of the world's inhabitants. This creates a fundamental problem that demands our thought.

Creating this thought will require moving beyond the traditional bounds of bioethics—with its focus on property, consent, privacy, and inclusion—to ask more fundamental questions about the power and meaning of novel advances in science and technology. Genomics is quickly becoming a powerful institution that produces routinized sets of practices that order human beings. It must take responsibility for this fact and support a thoughtful genomics that has the capacity to reflect upon and speak about these constitutional powers (Jasanoff 2011).

As Arendt suggested, creating this thought will require stepping back from that which is immediately present to create a gap from which to observe and reflect on the particularity of practices of knowing the world. Only then will it be possible to recognize differences in modes of knowing, and the meaning of those differences (Arendt, 1954 [2006], p. 9; Wolken, 2018).

Yet Arendt would not want us to stop at these philosophical musings. We think in places, in particular institutional contexts. We must create the interstitial places—the gaps— in these institutions that allow for reflective thought.

The Summer Internship for Indigenous Peoples in Genomics (SING) provides a critical example of how this might be done. Founded by genome scientists and indigenous scholars, SING trains indigenous scientists in genomics in a manner that makes space for reflection on the ethical, political, and cultural questions raised by the field. The goal of SING is to transform genomics so that it is responsive to these questions (Wade, 2018). For example, SING alumni have created new ethical guidelines that call on researchers to attend to concerns in indigenous communities about the meaning and effects of genomics (Claw et al., 2018). Concerns about discrimination must be taken seriously to create not just ethical science, but a reflective science that can understand the origins and effects of the categories it deploys. To achieve this, the guidelines urge scientists to work with indigenous communities early on in a research initiative to identify and respond to these and other concerns. This requires giving up on the idea that only genome scientists have relevant expertise about genomics. As genome scientist and SING collaborator Deborah Bolnick explains, '[Y]ou have to be willing to not see yourself as the authority, but rather as somebody who is going to listen to authorities' (Wade, 2018). Bolnick and other genome scientists taking part in SING are creating the conditions for the kind of public intelligence that Stengers called for that would—in Arendt's sense—create a thoughtful genomics.

For a thoughtful genomics to take hold, it must not just happen during the summer, but become an integral part of institutions. As the guidelines explain, commitments made by research agencies and universities to its ethical principles must be made manifest in 'supportive policies and substantial investments' (Ibid.) In other words, it is not enough to profess the principles. Institutions must change, devoting significant resources to creating legitimate, well-resourced spaces for more thoughtful—and, thus, more inclusive—science.

At my own university we have spent the last decade crafting this kind of space. Specifically, we created the Science and Justice Training Program (SJTP), which trains science and engineering graduate students alongside social science, humanities, and arts graduate students. Students learn how to speak to one another across their differences, and to recognize and respond to different modes of knowing (Reardon et al., 2015). This generates new forms of thoughtful practice that integrate questions of ethics and justice into disciplinary methods (Science and Justice Research Center Collaborations Group, 2013).

While our focus is on graduate student training, to make this possible we also have had to learn how as an institution to work across differences. For our natural scientists and engineers this has entailed finding a practical means (for example, in curriculum planning) for recognizing that reflective work counts as 'science,' and is not a distraction from the hard work of the lab. For our social scientists and humanists, it has meant stepping away from assumptions that as soon as you introduce a sequencing machine you might be producing a reductionist, determinist, unthoughtful vision of humanity.

Many hope for a future in which genomics and its offspring (i.e. personalized medicine) achieve their promises: new cures; longer and better lives. However, for that to come to pass, the field must support meaningful, thoughtful speech about that which it does. A response-able, thoughtful genomics should be the next milestone.

Acknowledgements

I would like to thank Michael Barilan, Margherita Brusa, and the reviewers at Oxford University Press for their comments and edits that sharpened the thinking and clarity of this essay. Many thanks to Joseph Fins for deepening my thoughts about Arendt and her concept of logicality. Finally, I offer my gratitude to the Pontifical Academy of Sciences for providing a beautiful and inspiring environment in which to think about genomics and personalized medicine.

17

Islamic *Sunni* perspectives on the ethics of precision medicine

Mehrunisha Suleman

Introduction: need for religious and socio-cultural perspectives

While the business of precision medicine is largely construed as a scientific endeavour, it is embedded within socio-political and ethical values and practices. The undertaking of precision medicine as an enterprise is an increasingly global one, where countries in the Middle East and Asia are rapidly joining efforts of North America and Europe to develop new technologies, including population-level genomics databases, whole-genome sequencing and association studies (Ghaly et al., 2016).

This chapter analyses what these values and commitments may be and how they may inform the design, implementation, and uptake of precision medicine in such milieus. Where pertinent, the chapter outlines any differences in perspectives that may exist between different locales and communities when deliberating moral commitments around precision and personalized medicine.

Theological, ethical, and historical accounts within the Islamic tradition as applied to precision medicine

Intentionality and teleology

As a normative tradition, Islam is comprised of its primary texts, the *Quran*, which is believed to be the word of God by Muslims, as well as the tradition of the Prophet Muhammad captured in authentic narrations (*Hadith pl. Ahadith*). Muslim scholars and lay people alike consider these two sources as

Mehrunisha Suleman, *Islamic* Sunni *perspectives on the ethics of precision medicine* In: *Can precision medicine be personal; Can personalized medicine be precise?*. Edited by: Yechiel Michael Barilan, Margherita Brusa, and Aaron Ciechanover, Oxford University Press. © Oxford University Press 2022. DOI: 10.1093/oso/9780198863465.003.0017

principal in guiding all aspects of sacred and mundane life. As the *Quran* and *Ahadith* do not offer distinctive guidance on emerging problems, religious experts are called on to analyse and interpret the textual sources through consensus (*Ijma*) and analogy (*Qiyas*) through the implementation of individual and/or collective reasoning (*Ijtihad*) to derive legal edicts (*Fatawah*). The Islamic tradition comprises two main branches, *Sunni* and *Shi'a*. The *Shi'i* school relies not on *Qiyas* but predominantly on *Ijtihad*. This normative divergence within Islam has implications for legal and theological perspectives. Due to constraints, this chapter is limited to perspectives from the *Sunni* school.[1]

Meta-ethical analyses of the Islamic tradition's engagement with biomedical advancements reveal tensions. Religious scholars find themselves ill equipped to traverse the aims, language, methods, and values of scientific enquiry in order to deliberate emerging ethical challenges on religious grounds. This has led to collaborative efforts between religious and scientific experts in what has been titled collective *Ijtihad* (*Al-ijtihad Al-jamai*).[2]

Despite such institutional challenges on the nature of Islam's engagement with biomedical science, the tradition is replete with values that Muslims adhere to and scholars and Muslim scientists rely on to guide their ethico-legal deliberations. Foundational to the Islamic worldview is the purpose of humankind's creation and how understandings of *telos* guide intentionality in each action. The *Quran* provides an account of the nature of humankind whose ultimate goals are the worship of God (51:56) and acting as God's representatives or deputies (*Khalifah*) on Earth (6:165). The latter privilege coexists with accountability as the aforementioned verse also states 'He may try you through what He has given you'.

Within the Islamic moral framework, humankind's capabilities (physical, emotional, and spiritual) are a trust or *Amanah*. How this *Amanah* is discharged influences one's ability to seek salvation, where every action in this earthly existence has metaphysical consequence in the everlasting life to come. In the *Sunni* tradition of Islam, the anchoring in otherworldly metaphysics does not align morality with naturalist ethics, but with God's expectations, as known through Revelation and Tradition (Padela, 2015). The Prophet Muhammad explained that actions follow intentions. Thus, within the Islamic moral framework, moral evaluation is focused on intention

---

[1] For more information on the Shi'i school please do refer to: Tremayne (2009) and Suleman (2020).
[2] For a detailed account on such scholarly meetings and deliberations, see Ghaly (2018).

(Sahih Bukhari Book 1, Hadith 1). Intentions ought to be rooted in worship or *Ibadah*, where outside of ritual worship one must align ones actions with God's will. An aid to determining God's will is a demarcation outlined in the Quran. Verses outline the dangers of corrupting the earth (*fasadan fil ardh*) (2:11–12); others instruct a stable balance that ensures poise or harmony (55:7–8).

Determining God's will has occupied theologians, ethicists, and legal experts across the ages. Deliberative instruments such as moral theology (*Usul al Fiqh*), legal maxims (*Qawa'id Al-Fiqh*), as well as the objectives of the law (*Maqasid al Shariah*) have been developed to enable scholars to deliberate the Islamic ethico-legal weight of a proposed intervention. Humankind's telos or encounter with God's judgement and God's balance (*Mizan*) in the afterlife is a focal point of all intentions and actions.

In relation to personalized medicine, all of this means that engagement and uptake of 'omics', require evaluation of the impact of such developments on humankind's purpose—worship (*ibadah*). If one's purpose of existence is worship, then one can argue that if developments in precision medicine help to serve this purpose then such endeavours are not only justified but ought to be encouraged. Proponents of personalized medicine may suggest that prevention of disease and/or effective treatments will mitigate an inability to offer prayers.[3] The *Quranic* narrative, however, presents an immediate challenge to this hypothesis, emphasizing that humankind is exceptional among God's creation. God describes in the Quran that 'We have indeed created man in the best of moulds' (95:4), meaning that we are endowed with capabilities required of us to discharge our religious worship. This fundamental verse holds true for all humans, even those who are sick and disabled. Rather, it provides a perspective of what is considered 'best' in God's eyes. This encompasses every form of human existence including that which is mired by disease and disability. Furthermore, the Quran states that God does not cause humankind to carry a burden or duty more than she or he can bear (2:286). Worship is co-dependent with capabilities. The infirm are either exempt or able to adapt ways of offering prayers (2:184; 9:91). Even what might appear an enhancement of capabilities for worship cannot be justified by the

[3] Performing the five daily prayers is considered the most emblematic form of worship in Islam, and distinguishes a believer. Sahih Bukhari Book 97, Hadith 159: 'A man asked the Prophet: 'What deeds are the best?' The Prophet said: 'To perform the (daily compulsory) prayers at their stated times . . .'. Other than the testimony of faith, the other three pillars are performed at fixed times in the year and not daily, i.e. fasting in Ramadhan, offering Zakaat, and performing Hajj.

paradigm of creation in the perfect mould. The teaching about the perfection of human nature as created by God entails the counter-intuitive presumptions that the Angels are more perfect than humans. Additionally, an argument to enhance capabilities of existing and future generations would also imply that previous generations, who were unable to benefit from enhancement medicine, are disadvantaged in ways that might interfere with God's attributes of mercy and intergenerational justice (Suleman, 2021).

Indeed, the Islamic theological and ethico-legal paradigm proposes an ontological reality traversing space and time that includes a primordial and post-corporeal existence. This ontological paradigm requires a nuanced and cautious reflection to fully assess precision medicine within Islam's cosmological framework.

Predictiveness, ontology, and prayer

That this physical earthly existence follows a primordial gathering of souls (7:172) emphasizes the *Quranic* message of worship, God's oneness, and humankind's purpose to bear witness to God. This witnessing is a form of being in constant worship while being God's steward. One might question whether, if not for worship or *Ibadah*, the undertaking of precision medicine could be justified or encouraged on the basis of enabling or enhancing humankind's purpose of stewardship? Envisaging ourselves as purely biological organisms, one can quickly derive justification for this. For example, one may consider optimization of such technologies within healthcare rendering humankind more able to exercise stewardship over creation. This simplification, however, overlooks the significance of disease and infirmity to humankind's spiritual existence.

Quranic and Prophetic sources distinguish between illness and disease, health and wellbeing. A disease may comprise a pathological state, yet illness constitutes spiritual disharmony that may or may not correspond with disease (Sahih Muslim Book 22, Hadith 133). Within the Islamic tradition, there is an inextricable link between the soul and the human body where each impacts the other (Ahmed and Suleman, 2018). Spiritual wellbeing is not solely dependent on physical capabilities; rather it is cultivated through prayer (Ibn Majah Book 37, Hadith 4244) and perfecting *Taqwa* or reliance on God. In this light, stewardship is a process of spiritual development through good action rather than mere discharge of responsibilities. Furthermore, accounts of Islamic theodicy propose that illness and disease may occur as a result of religious transgressions and that virtuous endurance is a means to expiate sins and uplift the devotee (Al-Shahri and Al-Khenaizan, 2005).

Some religious authorities suggest that a 'disease-free' existence may cause humankind to lose a critical connection to prayer, bringing about a recodification of suffering and spiritual development (Ahmed and Suleman, 2018). Others argue that despite such potential consequences, the Islamic ethico-legal framework gives priority to physical wellbeing. Al Aqeel, a contemporary academic and geneticist, states that: 'Physical wellbeing has precedence over religious wellbeing' (Al Aqeel, 2007, p. 1295). Such polarized perspectives play centre stage within the Islamic scholarly discourse on how to approach 'omics' and the potential that such advancements offer.

Contemporary professor of Islamic ethics, Muhammad Ghaly, explains that divergent views include a 'precaution-inclined approach' and the 'embracement-inclined approach' (Ghaly, 2018, pp. 47–49). The former advocates restraint. The opposing view calls for engagement with knowledge that can be of benefit even before science masters fuller knowledge of the enterprise in question and its impacts.

Ghaly provides an example of a scholarly view, presented here, challenging the normative and epistemic structure of scientific research, including that of precision medicine, as being oblivious of Islamic values and commitments. Ajil al-Nashmi, a religious scholar from Kuwait, advocates critical theological evaluation of biomedical advancements, stating that 'many of the results of modern scientific research are not in conformity with the *Shariah* because the leaders in these fields are not guided by religious values and are mainly motivated by material interest and personal desire' (Ghaly, 2018, p. 60). He also offers the distinct view that, pertaining to genomics, all actions are by default prohibited (Ghaly, 2018, pp. 47–79). His arguments cohere with the Islamic tradition's emphasis on intentions explored above, which would state that if the industry of science is motivated by avarice and pride, then by default its products are immoral.

The notion of benefit

From the biomedical perspective we surmise that the vast repositories of individual genomic, metabolomic, and proteomic samples will herald unprecedented knowledge, analytic capabilities and subsequent prognostication. Against this, Muslim scholar, academic, and physician Professor Aasim Padela maintains that this prevailing biomedical discourse has reduced human beings to: data stores that house information; reproductive organisms; and an evolving biological entity (Padela, 2018). The role of humans as walking repositories of data is of predominant interest to the industry of precision medicine and the promise of predictive technologies to offer prevention and treatment.

Precision medicine's promises in terms of prediction and prognostication directly challenge Islamic theological understandings that *Qadr* (fate) is God's purview. Seemingly, biotechnology provides humankind with unprecedented capabilities of altering what may have been 'written' or predetermined in the genetic code. Al-Nashimi warns against misconceptions regarding the nature and source of predictive knowledge as well as against the rise of purely scientific worldviews and its eclipse of spiritual and moral anthropologies. People's rush to embrace omics-based modalities of care might result in overdiagnosis and medicalization, thus exacerbating the erosion of the spiritual and religious ontology.

Furthermore, the preoccupation with the predictive potential of such technologies may also detract from our understandings of human potential, responsibility, and accountability. As communities and societies, we are enriched by differences, both in our physical and spiritual potential. Biotechnology might push humans towards less diversity and towards conformity with 'best' laboratory values and similar markers. Excessive preoccupation with optimization of worldly 'wellbeing' and unbridled search for longevity go against Islamic teachings about the After Life and God's missions for human kind. The Holy *Quran* says: 'and what is the worldly life except the enjoyment of delusion' (57:20). 'But you prefer the worldly life; although the life to come is better and more enduring' (87:16–17).

Amanah, ownership, privacy

A vital moral concept within the Islamic tradition is *Amanah* or trust. A key aspect of this trust is custodianship over our own bodies. Hence, from a theological perspective it is imperative to inquire whether 'big data' science and services related to the acquisition, ownership, storage, and use of genetic and other omics derived from human bodies are compatible with this stewardship.

In terms of acquisition or giving of samples and tissue, there is a view that 'our bodies belong to God' (Hamdy, 2012) such that individuals may not hand over any aspects of themselves as 'to God we belong and to Him we return' (2:156). Such beliefs also explain the lack of distinction made within many Islamic ethico-legal narratives between the sanctity of the living and dead body. Some may extend such concerns to the acquisition of samples during a body's life or death (Mavani, 2014).

However, it is also argued that a person's stewardship over his or her body entails a responsibility to gain benefit for that body. Additionally, the relief of sickness is a good in its own right: 'indeed God did not make a disease but He made a cure for it' (Tirmidhi, Vol. 4, Book 2, Hadith 2038). Such understandings thus provide justification from an Islamic moral framework for the obtaining, use, and storage of samples necessary for the undertaking of precision medicine.

The ownership of individual samples and the derived data or knowledge is contentious, as, from the above, it is evident that from an Islamic standpoint the material would either be owned by God alone, or through stewardship, by the individual from whom the sample originated. In any case, the ownership could never be handed over to or appropriated by a third party, including a government, company, or industry. The practical implication is that any use of the samples ought to require proper informed consent by the patient. A distinction must be drawn here between 'omics' data storage and use for population-level research and that for patient/familial diagnosis and treatment. Reservations on ownership would apply for the former but not so for the latter. Ghaly and colleagues, for example, have written about the need for disclosure of genetic information in the case of incidental findings for children and other family members (Ghaly et al., 2016). More work also needs to be done within the Islamic ethico-legal discourse in delineating the difference between anonymized and identifiable genomics data and whether that alters ownership and subsequently stewardship.

Islamic moral understandings of *Amanah* point to decision-making and discharging trust that is not limited to us as individuals or as a current generation but to ensure intergenerational justice. The allure of short-term progress means we must be wary of the potential negative and unintended consequences of such large-scale data acquisition including its sale to and/ or unauthorized access by third parties (8:27). Contemporary Professor of Islamic Bioethics, Abdul Aziz Sachedina, points to the Islamic principle of 'prevention of harm being preferred to doing good' (Sachedina, 1998). The collection, storage, and use of data also raises challenges of privacy. Islam emphasizes the virtue of privacy and guarding the secrets of others (Tavaokkoli et al., 2015). Commercial interests might create misuse of data and unintended prejudices against individuals. These outcomes are counter to Islam's teachings on the custody over the human body.

Policies, guidelines, views, and practices from the Muslim world on precision medicine

Burgeoning investments in genomics and precision medicine, particularly in the Gulf region (Gulf Cooperation Council), may result from the high prevalence of genetic diseases there, mainly due to consanguinity (El-Hazmi, 2003; El Shanti et al., 2015). The Muslim world looks upon the Gulf states as pioneers in religious and moral deliberation in relation to technology and society. The *Maqasid* principle of 'protection of progeny' may also be a driver for such investments and research (Banu az-Zubair, 2007). Governments in the region have launched ambitious initiatives to kick-start efforts towards 'omics' research. Such programmes include the Centre for Arab Genomic Studies[4] and the UAE Human Genome Project in the United Arab Emirates, the Arab Human Genome Project and Saudi Human Genome Project[5] in Saudi Arabia, and the Qatar Genome Project (Qoronfleh and Dewik, 2017). Some of these projects have sought to engage religious authorities and scholarship, to ascertain Islamic ethico-legal perspectives on emerging challenges pertaining to 'omics' (Ghaly et al., 2016).

There have also been regional and national policy and regulatory efforts to evaluate population profiles in the region as well as an appraisal of religious and cultural values pertinent to the implementation of precision medicine technologies (UNESCO Cairo Office, 2011; Ghaly et al., 2016). Except for a detailed Islamic ethico-legal analysis around incidental findings, the report provides little theoretical and practical guidance on issues highlighted in the previous section, including privacy, ownership, intentionality and predictiveness. Despite its narrow scope, the report importantly calls for further research in Islamic ethics and for scholarly deliberations to engage widely with the industry or genomics and precision medicine in the region. The authors also recommend the importance of such religio-ethical engagement being placed within wider global bioethics discourses. The authorship includes key stakeholders in the region. The Qatar Genome Project, for example, has also embarked on a public engagement initiative—seeking to

---

[4] www.cags.org.ae
[5] https://shgp.kacst.edu.sa/index.en.html (last accessed 27th October 2021).

consult the Qatari public on their views about the potential use and impact of genomics technologies. This initiative is the only one of its kind in the region and throughout the Muslim world (Nun et al., 2019). It is likely to yield important insights into the views of lay Muslims on the nature, implementation, acceptance, and understandings of such advancements. Efforts to embed religio-spiritual concerns within policy deliberations is key to ensuring that biomedical progress in the region coheres with the moral universe of Muslims. The principle of *Shurah* or consultation is key within the Islamic tradition. It is unclear whether this is the driver for such public engagement initiatives or whether it is due to a commitment to align with emerging global standards.

Although the focus in this chapter has been on the overall industry of 'omics' and a consideration of the aims and impact of the broader project, socio-cultural research shows that religio-spiritual commitments are pertinent at the 'personalized' end of the endeavour, such as spiritual commitments on prevention and treatment strategies (Yeary et al., 2019). The discourse on precision medicine goes much beyond regulation and public policy. Many Muslim patients and families seek rigorous guidance on the arrival of novel biotechnologies at their bedside. The laity rely on faith to arrive at meaning-making around illness and suffering as well as the nature of being and existence and the role and importance of God as Creator, Sustainer, Healer, and Source of all knowledge. The interplay of faith and biomedical advancement as well as religious and political authority in such contexts will provide for rich socio-anthropological study.

Adl and *Ihsaan*

Despite the promised benefits of governments and scientists in countries investing in 'omics' technologies in the Muslim world, there is a critical limitation to the potential benefit of these industries to populations in the region. Currently limited to indigenous populations alone, migrant subgroups—particularly migrant workers and displaced communities whose occupational and other health needs are emergent and require careful planning and provision—are not listed as potential beneficiaries of such initiatives. Quranic principles of *Adl* and *Ihsaan* (16:90) call towards ensuring 'justice' and 'goodness' for all. This duty relates to the Prophetic description of the Muslim community as one body (Sahih Muslim Book 45, Hadith 84) and to the teaching that nations and races are all equal, none having right or status above another

(Prophet's Last Sermon) (Acheoah and Hamzah, 2015). One potential harm of precision medicine leading to further global health inequalities is a risk that such regions will need to address. The scale of investment in such technologies also points to a potential shortcoming in ensuring basic healthcare for all, including affordable public health for migrant workers and displaced communities (Adhikary et al., 2011; Joshi et al., 2011).

Conclusions

This chapter has offered an overview of the theological, ethical and historical accounts within the Islamic tradition as applied to precision medicine as a concept and as a growing force in healthcare and science. It suggests that progress in 'omics' presents a challenge to the institution of Islam and in certain religious scholarly discourses. Such deliberations will need to engage with theological and moral concerns around *Amanah*, privacy, ownership, and intentionality; salient analyses of which have been presented here. Muslims rely on faith values for meaning-making around illness and are likely to seek deliberative forums and resources to assess the acceptability of what precision medicine has to offer. Finally, Islamic moral teachings around *Adl* are likely to inform theoretical and practical discourses on the impact of precision medicine being able to mitigate or exacerbate inequalities in health, both locally and globally throughout the Muslim world.

Acknowledgements

The author would like to acknowledge and thank Sunnah.com from which all references to Hadith have been obtained for the chapter. The author would also like to acknowledge and thank Corpus Quran, where, unless otherwise stated in the chapter, from which the Quranic verses have been taken: http:// corpus.quran.com

18

Genetics, genetic profiles, and Jewish law

Yehoshua Weissinger and Yechiel Michael Barilan

Jewish ethics and Jewish positive law

Like other medical disciplines, genetics has been making fast progress in search for better diagnoses and treatments, as well as by its growing impact on metaphysical, moral, and social discourses. The advent of germ theory, antiseptics, and antibiotics also changed scientific knowledge and medical practice, and no less altered perceptions of health, illness, and social relationships. Germ theory revised notions of 'infection', 'purity', and 'contagion'. Genetics has been challenging notions of identity, self, and kin relations. These topics are of utmost relevance to Judaism and other world religions that are clan-based and in which kinship is a central value. However, because Jewish law follows the paradigm of positive law more than natural morality, it is quite difficult to specify from Jewish law to genomics. Principally, Jewish law beholds as licit any act that is not prohibited by the ancient laws of the Talmud (the authoritative source of Jewish law, dated to late antiquity). With the absence of Talmudic and other ancient authoritative references to genetic knowledge and treatments, they are *prima facie* acceptable (Barilan, 2014, introduction). Indeed, because there are no old laws against germline editing, the collection of personal omics, and even human cloning, Jewish law is *prima facie* tolerant of these practices (Barilan, 2005).

Even though genomics and information technology did not exist when Jewish law was consolidated, Judaism has always beheld critically attempts to tell the future and to decipher the human personality. Both endeavours are relevant to personalized medicine—its aspiration to predict the personal health of every person.

The rabbinic sources teach that man is partner to God in the act of creation. Everything needs perfection. The perfection of nature is humanity's task, to be realized by means of rationality. Hence, there is a special value in employing science and technology in the service of medicine and the public

Yehoshua Weissinger and Yechiel Michael Barilan, *Genetics, genetic profiles, and Jewish law* In: *Can precision medicine be personal; Can personalized medicine be precise?*. Edited by: Yechiel Michael Barilan, Margherita Brusa, and Aaron Ciechanover, Oxford University Press. © Oxford University Press 2022. DOI: 10.1093/oso/9780198863465.003.0018

good. In this light, the rabbis behold genetics and genomics as *prima facie* good pursuits.

However, Jewish law warns against 'licit vices', which are acts that are unvirtuous or harmful even if they do not violate any formal law. The law authorizes the rabbinic leadership to intervene and introduce temporary legislation whenever they deem it necessary for the 'reinforcement of religion and the good order of society' (Maimonides, *Mamrim* A:2; *Maimonides's Introduction to the Mishna*, ch. 4).

In his gloss on the Pentateuch, the medieval Rabbi Nachmanides interprets the Biblical decree 'Ye shall be Holy, for I the lord Your God am Holy' (Leviticus 19:1) in this light, saying that unvirtuous, opportunistic conduct within the formal limits of the law is the epitome of unholiness. In a way, the 'meta-legal' discourse in Judaism is the arena where ethics, politics, and especially theology play out their influence on Jewish normative conduct.

In broader terms, contemporary experts in Jewish religious law (*Halakha*) distinguish between 'the law' and 'meta-legal' factors (Cherlo, 2002). The latter addresses a broad array of considerations, such as virtue ethics, sociopolitical implications, and the image of Judaism in the eyes of Jews and non-Jews.

Consequently, even though Jewish law says nothing directly about genetic profiling and its use in personal decisions about marriage, career, childrearing and hobbies, the Jewish tradition at large can offer us ways to consider whether such practices might be considered 'vicious', and how genomics may develop in light of the values promoted by Judaism.

This chapter aims at the explication of Jewish moral anthropology in relation to genetics (a branch of science) and genomics (a branch of applied and socially embedded set of technologies). Because these norms are not integral to Jewish law, they are liable to change, as socio-historical circumstances change to accept more readily the personality and teachings of the rabbis.

Personal and social profiling

Forty years ago, Victor Barnouw defined 'personality' as a 'more or less enduring organization of forces with the individual associated with complex of fairly consistent attitudes, values and modes of perception, which accounts in part for the individual's consistency of behavior' (Barnouw, 1985, p. 8). Knowledge about a person's 'personality' comes from observation of his or her behaviour in search of three main patterns—*traits*, which are distinct behavioural regularities, *character*, which refers to interpersonal dispositions and conducts, and *modes of organization* by which the person tends to experience reality and to represent it (Bok, 1988, p. 42).[1]

This paradigm of the human personality, formulated on the eve of the Human Genome Project, and much before contemporary behavioural genetics emerged, already contains the element that fit the geneticized discourse of 'personality'. There is a supposition of 'consistency' of behaviour and mental life (cognition and emotion), which is anchored *within the person/body*. However, when people talk about 'personality', they refer to a person's actual conduct; when geneticists talk about 'personality', they refer to statistical values whose accuracy correlates with the size of population sampled. Behavioural genetics does not tell us whether a person is shy or depressed but gives a number that indicates the chances that a person is shy or depressed. Moreover, genetic information can influence behaviour and perhaps even personality. If a person is told that he or she is 'genetically shy', he or she might behave accordingly and incorporate 'shyness' in his or her sense of self. The placebo/nocebo effect of genetic information has been established by showing that 'merely receiving genetic risk information changed individual cardiorespiratory physiology' (Turnwald et al., 2019).

Key to the Jewish approach to genetic knowledge are the old sources on the nature of 'knowledge' in relation to human nature, especially the nature or personality of individual human beings.

Self-knowledge and self-search in Judaism

Whereas in the classical world self-knowledge is a fundamental need that is fulfilled by wise reflection and by consultation with oracles and similarly supernatural powers, the rabbis downplay introspection on fate and personal predictions. The Talmud points at three references for reflection as necessary for the good life. Two are generic, even truisms of human existence; one is about one's responsibilities:

> Know from where you came, where you are going, and before whom you are destined to give a judgement and accounting. From where you came—from a putrid drop; where you are going—to a place of dust, maggots, and

---

[1] We follow Gordon Allport's approach which is based on patterns and correlations among kinds of circumstances and patterns of behaviour (Allport, 1965, pp. 341ff; Hunt, 1993, ch. 11).

worms; and before whom you are destined to give a judgement and accounting—before the supreme King of Kings (Mishna, Avot 3:1 see also
Avot 2:1).

According to the Talmudic sources, certain kinds of inquiry for knowledge
are beyond human reach and even prohibited (e.g. Talmud Hagigah 13a).
The Bible says,

> As you do not know what is the way of the wind,
>   Or how the bones grow in the womb of her who is with child,
>   So you do not know the works of God who makes everything.
> (Ecclesiastes 11:5)

The Talmudic sage Jonathan ben Uziel writes, 'As much as you do not know
how the soul fills the fetus in the womb, and its sex [until birth], you will
similarly not be able to know God's creative wisdom' (Gloss Jonathan on
Ecclesiastes 11:5). The Talmudic literature does not rule out the existence of
extraordinary means to gain some useful information. A typical example is
King Saul's resort to the necromancer from Ein Dor (I Samuel 28:8–14). Even
though the necromancer made Saul see his impending death, the Talmudic
sources indicate that even when successful, modes of supranatural inquiry
yield information which is partial, distorted, misleading, even risky (Talmud
Hagigah 14b).

Whereas Saul violated an explicit Biblical prohibition on recourse to
witchcraft, the rabbis have considered the general precept 'Thou shalt be
perfect with the Lord thy God' (Deuteronomy 18:13) as a general instruction to avoid inquiries into areas of knowledge not considered proper for humans—the mysteries of Godhead, the mysteries of creation, fortune-telling.
Interestingly, the Hebrew word *tamim*, which the King James Bible rendered
'perfect', means 'wholesome', 'impeccable', and 'innocent'. The implication is
that human wholesomeness entails imperfect self-knowledge, the existence
of a 'blind spot' in relation to self and God. Partial knowledge could be more
truthful and less distorted than fuller knowledge in terms of the accumulation of data and its representation as detached, objective information.

This is especially relevant to close, interpersonal relationship. When the
Bible says that 'Now Adam knew Eve his wife' (Genesis 4:1), the knowing is
not an objective tally of physical traits; it is not information produced and
sifted through the mediation of technology. However, the use of the word
'know' in this context is neither metaphorical, nor misleading. Because even

when it lacks plenty of facts that could be gained by research and surveillance, intimate human acquaintance yields familiarity that is *wholesome*. It is 'perfect', even if it is not readily shareable with others, and even when it misses pieces of information. It is similar to mimetic knowledge (Wulf, 2008). However, it is not about emulation of a skilful person at work, nor is it learning from a role model, but knowing the other person as he or she is, while being one's own self in a relationship of intimacy and trust.

In his Gloss on the Pentateuch, the prominent medieval rabbi, Rashi wrote, 'Conduct yourself with God as *tamim*, without trying to know the future, but accept with wholesome attitude everything that happens' (Rashi, Deuteronomy 18:13). In this spirit the rabbis invoke the Biblical phrase 'God protects the *tamim*'. The nineteenth century rabbi Hirsch writes, 'God protects the simpleton who acts as ordinary people do, even if his conduct is somehow risky' (Exegesis to Deuteronomy 18:13).

A *Midrash* (Talmudic lore) tells a story of a man who asked King Solomon to teach him the language of the animals. The wise king tried to talk the man out of the idea, but ultimately granted his request. One day the man overheard animals discuss the future. But the man's preventive action backlashed (Zilberstein, 2000, p. 236; Hisdai, 1824). The message in this story is that it is sometimes better for a person not to know things about oneself, even when such knowledge is attainable by special skills. This story illustrates that certain kinds of knowledge are better avoided, even when they are not associated with witchcraft and the breaking of Biblical taboos.

From this survey follows a distinction between Judaism's celebration of applied technologies and Judaism's reservation with technologies of knowledge, especially in relation to the mysteries of self and the future. Although it is impossible to project from stories on animals discussing future events to real-life situations, we can behold genetics as problematized knowledge. Contemporary culture represents the genome as 'the book of life', and personal genetic variants as special knowledge about the self, the future self.

In 1973, the prominent authority in Jewish law Rabbi Moshe Feinstein (1895–1986) discussed pre-marital screening for Tay–Sachs, which is an incurable and fatal genetic disease of infancy, relatively common among Ashkenazi Jews. Principally, Feinstein explains, it is better not to test. The righteous conduct themselves wholly, not seeking future hidden problems. However, Feinstein says that once the exams became accessible and highly reliable, refusal to self-test is like turning a blind eye on danger that might bring much suffering to others. He concludes that self-testing is desirable, but only when a person starts contemplating marriage (Feinstein, 1985,

p. 21). Feinstein warns at length against the risks of psychological harm and stigmatization brought about by such testing.

It seems that according to Feinstein, the moral focus is not on the knowledge in question, but the efforts made to obtain it. In the spirit of the medieval distinction between 'ordinary' and 'extraordinary' means, it may be suggested that whereas God expects people to exercise ordinary measures of precaution, extraordinary efforts to foretell personal events are unvirtuous, not expressions of supererogatory respect for life.

Three points of consideration might be at stake. The first is the scientific validation of the technology of knowledge in relation to specific events. Non-evidence-based methods, ambiguous information, and mere statistical data (e.g. risk level of 30%) are a source of harm and confusion. Second, high-end rarely used technologies are considered beyond human duty and virtue and may better be avoided, especially if they do not lead to preventive actions backed by expert consensus, such as professional guidelines. The third consideration is the ultimate nature of the *personal* information sought. The less it is related to protect life and prevent disease, the less it is desirable to pursue. Information about beauty, talent, and virtue are to be avoided. Such traits are for direct interpersonal observation within a proper social context.

These three considerations are usually interconnected. Genetic testing for 'musical talent', for example, is not scientifically validated; it requires sophisticated technology and has nothing to do with the values of health and life. Because the accessibility factor is the one most rapidly changing, the plummeting costs of testing for 'talent for music' might tempt many people to seek such exams, to consider them 'ordinary means' of good parenting. However, as Rabbi Feinstein indicates, the social cost of prejudice and stigma, especially when directed at choice of mate, may be no less significant than pecuniary costs.

Beyond genetics—human responsibility

The Talmud narrates a dialogue between the Prophet Isaiah and King Hezekiah. The virtuous king refrained from having children following a discovery through spiritual signs that he would beget an evil son. The Prophet rebuked the King for acting on personal prophecies that were beyond the reach of ordinary humans. He instructed King Hezekiah to fulfil his duty to procreate and to dedicate efforts to the education of his children. 'Do what humans are supposed to do' he told him, meaning that humans should raise children and educate them well, not seek knowledge about the child's future (Berakhot 10a).

Because the Talmud emphasizes the piety of the King earlier in the discussion, it is evident that the prophetic insight was not sinful. It might have been an 'accidental finding' coming to the King by means of the Holy Ghost. However, deviance from human responsibility that is not related to the protection of life is clearly discouraged. Hezekiah's righteousness and efforts notwithstanding, his son Menashe became Israel's wickedest king. However, the Talmud does not consider this tragic outcome as a retrospective justification of the King's decision not to have children. Earlier in the Talmudic discussion appears a story of a rabbi who suffered from evil neighbours and prayed for their death. His wife told him that he should pray for the demise of sin, not sinners.

It seems that the Talmud prefers attention to behaviour, which is under human control, at the expense of attention to essence, which is not. From this perspective the potential harm from search into hidden traits and future events is distraction from, and weakening of, human striving, a shift from responsibility to fate. The Talmud contains a few stories about people who defied their fate as forecasted by astrology and similar means (Shabat 156a–b). Moreover, the Talmud contains advice on how to overcome one's fate or sublimate it. For example, a man who was born under the sign of bloodshed is not destined to be a mercenary. He can work in an abattoir.

It follows that, when *life and health* are at stake, attempts to predict future events are virtuous, provided they pursue validated and rational methods. However, when other values are in question, humans had better act 'naively', avoid questioning, and cope with developments adaptively. This appears to exclude the use of genomics and other modalities in the service of enhancement medicine. However, whereas secular bioethics regards technologies of intervention (e.g. manipulation of the genome) with greater caution than technologies of knowledge (e.g. genetic testing), rabbinic law and ethics seem to be more guarded with the latter, and less restrictive with the former.

Profiling personalities

Libel (*Lashon Hara*) is one of the most vicious acts in Jewish law and ethics. In Jewish law, libel is the spread of harmful information about a person, even when truthful, and even when not harmful. The Talmudic sages construed libel as one of the worst vices; Jewish moral literature has stressed the virtue of distancing oneself as remote as possible from either speaking or hearing libel (Kagan, 2008, p. 12).

Culpability

In Judaism, the notion of libel highlights information about persons and groups of persons as morally laden, and infinitely so, even if no malice is involved.

However, there is a duty to inform a potential victim of an imminent harm. For example, if a person carrying an infectious disease is walking freely in the neighbourhood, there is a duty to announce it (Wilner, 1973, Mark 43; Margaliot, 2007, Mark 573).

Ultra-orthodox Jews tend to gather information about eligible bachelors prior to dating. Financial capacity of the family and personal career options of the candidates are typical topics to be discussed in advance. Many wish to inquire about health problems. Jewish law encourages the disclosure of a few serious health issues, such as epilepsy (Waldenberg, 1985, Mark 4). Because marriage has a contractual aspect, non-disclosure of certain personal information may result in nullification of marriage. Health problems of this kind must be disclosed in advance (Margaliot, 2007, Mark 507).

Contemporary rabbis have ruled that the Talmudic and medieval list of disabilities that can nullify marriage is not fixed, but it changes with the transformations of scientific knowledge and cultural norms. Some examples given in the sixteenth century codex of law *Shulhan Arukh* bring into relief the subjective and culturally constructed aspect of some such issues, for example, a scar from a dog bite located in a private part of the body (*Even HaEzer* 39:5).

Contemporary rabbis stress openness and honesty as essential for successful marriage. However, early on, before affection has set in, people may wish to know about genetic risks and other health-related issues a person might have tested for. A culture of testing and profiling might set in, sending young people who have not tested (and cannot hide any results) to test and to disclose.

Testing for all sorts of risks and personality traits invites the problem of disclosure. First, hiding something from a spouse is considered unvirtuous (Waldenberg, 1988, Mark 48). Second, if the risk or trait are trivial, it is difficult to explain the person's motivation to self-test. It follows that taking genetic tests pushes people to the brink of a slippery slope of confusion and stigmatization. Many people know that genetic markers provide partial and uncertain indication of future events and that social circumstances, personal choice, and mere chance are no less determinant of personal traits and actions. However, because it is possible to reify worry as scientifically validated risk, and because personal, social, and unknown factors are harder to reify, genetic prediction is likely to induce both confusion and stigmatization.

The 'Golden Rule' requires people to disclose information they would not wish to be withheld from them. If the social standard is awareness of certain risks, non-disclosure is a kind of cheating. It follows that testing for well-established, high-risk, serious conditions (e.g. Tay–Sachs) is expected of people; medically irrelevant, low-risk, or scientifically unfounded testing is discouraged. There might be a tier of tests which Jewish law wishes to discourage, but if they become a social norm, disclosure is expected. There might be a gap between the *theological* virtue of 'wholesomeness'—trust in God without questions, and the *interpersonal duty* to respect the Golden Rule. Even if the person does not care about the genetic information in question, he or she does have certain expectations regarding transparency and communication in intimate relationships.

Conclusion

Judaism and Jewish law have been reluctant to regulate science and technology. Additionally, as a system of positive law, even rabbis who are interested to set limits on technology find it quite difficult to extrapolate from ancient religious law to modern contexts. In contrast to Jewish openness to free scientific inquiry, technological innovation, and free markets, in Judaism *personal* information is a locus of legal control and moral caution. The Jewish tradition teaches us that forecasting *personal* future is fraught with error and bias, and that whole knowledge of a person as a person may be compatible with lack of knowledge about certain aspects of the person as a biological creature.

Jewish law on libel and the contracts of marriage teach us that personal information, even genetic information, is never a 'pure' scientific fact, but a social token. The rabbis have encouraged people to adhere to programmes of preventive medicine and to perform tests that are scientifically valid as necessary preventive measures of serious disease.

Regarding other predictions, even if aimed at important issues such as a child's future tendency to commit crimes, the rabbis advise 'not-knowing'. Education and intimate relationships are human endeavours and should remain so, even if it seems that biological predictions might improve the outcome.

Because personalized medicine is information-centred, the ethical discourse surrounding it might benefit from the Jewish sensitivity to personal information about future risks and outcomes—the desire to obtain such information, the means deployed, and the social role such information possess in society.

19

Pharmaceutical contributions to personalized medicine

Roger Perlmutter

Introduction

The phrase 'personalized medicine' is understood to describe a process wherein an understanding of pathophysiology permits adjustment of therapy for each individual patient, directed by consideration of that patient's unique situation. This individualized approach has characterized medical practice for millennia, but has been bolstered most recently by the appearance of sophisticated measuring tools that can inventory hundreds of physiological parameters, and by the development of computational algorithms that enable the interpretation of these measurements to improve both the diagnosis and the treatment of disease. Personalized medicine is thus sometimes referred to as 'precision medicine', reflecting these advances.

Whereas personalization, in this sense, seems straightforward and desirable, in the current context we understand that the practice of personalized medicine rests upon the view that subtle changes in standard indices, including measurements of physiological performance across a range of domains as well as cellular and molecular characterization of biological fluids, will detect asymptomatic disease, or even predisposition towards a particular ailment, at a time when preventive measures can be fruitfully employed. The assumption underlying this formulation is that a highly precise measurement system can yield data sets of sufficient size and robustness to permit a personalized assessment of the risk of illness. Phrased in this way, the availability of large data sets has predictive value. In some circumstances, knowledge of a predisposition to disease that is, at the time of measurement, asymptomatic should permit the introduction of countermeasures that will reduce the risk of development of the disease itself. Detailed evaluation of individual variation thus permits the development of medical practice that is personalized,

Roger Perlmutter, *Pharmaceutical contributions to personalized medicine* In: *Can precision medicine be personal; Can personalized medicine be precise?*. Edited by: Yechiel Michael Barilan, Margherita Brusa, and Aaron Ciechanover, Oxford University Press. © Oxford University Press 2022. DOI: 10.1093/oso/9780198863465.003.0019

predictive, preventive, and participatory—so-called 'P4 medicine' (Flores et al., 2013).

Modern devices permit evaluation of health status in exceptional detail, using imaging devices, wearable instruments that record physiological performance, and molecular and cellular analyses of biological samples, including blood, urine, and stool. These measurements can then be compared with baseline parameters deduced from the evaluation of millions of patient samples. An important feature of these measurements is that they also permit personalized, i.e. 'n of 1', experiments based on the goal of engineering a reversion to prior values ('normalcy'). The hypothesis in this case is that placing all measurable parameters within an interval of normalcy can help to ensure long-term health and wellbeing.

Ordinary medical practice, however, has less lofty ambitions: to alleviate suffering and enhance survival. It is understood that the appearance of grievous illness, fatal over some time interval, is inevitable for every individual. Hence medical practice seeks to extend and improve the lives of the grievously ill. This conceptualization places the *patient* at the centre of focus.

People become patients because they articulate a chief complaint or have an abnormality detected, such as high cholesterol level, one that a medical practitioner can try to address. As compared with P4 medical practice, the more modest goal of the physician/scientist is the *regularization* of human life span: we hope for a world where all children are born healthy, all transit adolescence successfully, most achieve some measure of satisfaction in their adult lives, and all complete a full, natural life span without undue suffering. That such a goal can be articulated is owed to the enormous progress made in developing new medicines during the past century. Advances in biomedical research have yielded a priceless portfolio of treatments that cure infection, alleviate pain, eliminate or delay heart disease, and reduce the burden of malignancy, in many cases permitting meaningful life extension. As might be expected, these advances have also revealed the underlying heterogeneity in disease processes previously thought to be uniform, thus encouraging greater specialization of practitioners and complicating both diagnostic evaluation and therapeutic planning in clinical settings. This too is 'personalized' medicine.

Finally, improvements in medical practice have altered the demographics of human populations in the world today, and will fundamentally change the character of human societies in the future. To place these ideas in perspective, one must first characterize the impact of drug discovery on life expectancy, especially in the adult years. In this article I will explicate the process whereby new medicines are introduced and cite numerous examples of the impact of new medicines on common, life-threatening diseases.

Modern medicines are improving life expectancy

Life expectancy from birth has improved in the main from beneficial changes in standards of living: improved sanitation and enhanced nutrition have contributed the majority of benefit throughout human history (Pruss et al., 2002). Since the mid-twentieth century, however, the most important improvements in human life expectancy from birth have resulted from the development of medical interventions. During this period, life expectancy improved by nearly two years per decade (Mehta et al., 2020), with much of this improvement secondary to the declining impact of cardiovascular disease as clearly shown in Figure 19.1. While some of this improvement resulted from a decline in smoking and other public health measures, the bulk of this effect was related to medical interventions and in particular to the control of hypertension, achieved through a variety of pharmaceutical interventions, and to the introduction of effective cholesterol-lowering agents. The crucial importance of cardiovascular disease in determining life expectancy

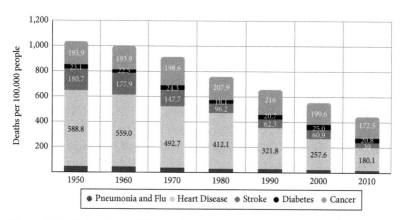

Figure 19.1 Reduction in US mortality from specific diseases between 1950 and 2010. Deaths per 100,000 people for major causes of illness are shown. Note the steep reduction in cardiovascular disease, which is still the greatest single cause of death in the USA.

in developed societies is illustrated by the relative stagnation of this metric since 2010, resulting from the lack of broadly adopted new interventions in the cardiovascular area (Mehta et al., 2020).

But it is not only in the domain of cardiovascular disease that modern pharmaceutical development has had an important impact. Paediatric vaccines and the development of effective treatments for infectious diseases have of course had a salutary effect on human life (Rappuoli et al., 2014). More recently, pharmaceutical companies have developed both prophylactic and therapeutic measures that are reducing the impact of cancer as well. Consider, for example, the effect of the introduction of an adolescent vaccine against human papillomavirus infection, a sexually transmitted disease responsible for most cases of cervical cancer in women, and nearly half of all cases of head and neck cancer in both men and women. One recent analysis conducted by the US Centers for Disease Control showed that introduction of the four-valent human papillomavirus vaccine resulted in a 64% reduction in papillomavirus infection caused by relevant strains in the period from 2009 to 2012, as opposed to what was seen in the period from 2003 to 2006, the year that the vaccine was registered (Markowitz et al., 2016). A recent study modelling the impact of the more recently introduced nine-valent papillomavirus vaccine in low- and low/middle-income countries predicted that vaccination would reduce the incidence of cervical cancer by 86.2–90.1% and avert 61 million cases during the next century (Brisson et al., 2020). Indeed, these studies indicate that cervical cancer could be *eliminated* in most low- and low/middle-income countries by the end of this century were uniform papilloma virus vaccination of adolescent girls adopted, a stunning example of the effect of advances in vaccine research on an important cause of morbidity and mortality in young women.

Beyond vaccination, during the past two decades important advances have been made in the pharmaceutical treatment of a variety of malignancies. Progress has recently accelerated with the advent of immunomodulatory therapies that stimulate elimination of cancer cells by the patient's own immune system. Figure 19.2 illustrates the result of treatment of a 72-year-old man with advanced melanoma that had metastasized to the patient's left lung and mediastinum. This patient had received multiple prior treatments with both high-dose chemotherapy and the immunomodulators interleukin-2 and ipilimumab, without success. Treatment with pembrolizumab, an immune checkpoint inhibitor, resulted in near-complete regression of all tumour masses, with rapid clinical improvement. These unprecedented responses in patients with previously untreatable malignant melanoma have

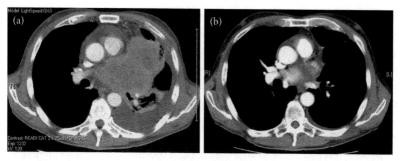

Figure 19.2 Computed tomographic images from a patient with advanced metastatic melanoma. *Left panel*: cross-section just below the aortic arch demonstrating large tumour masses in the left lung and adjacent mediastinum. *Right panel*: image taken 12 weeks after initiation of therapy with pembrolizumab by intravenous infusion at 2 mg per kilogram body weight every three weeks. Substantial clearing of the affected areas can be seen.
Images courtesy of Dr Antonio Ribas, University of California at San Francisco.

become commonplace and are observed even in the elderly population (Queirolo et al., 2019). For example, in August 2015 former US president Jimmy Carter announced that he had developed malignant melanoma that had metastasized to his liver and his brain. At that time, less than six years ago, patients with this degree of tumour involvement had remaining life expectancies measured in weeks. Just three months later, following treatment with pembrolizumab coupled with radiation administered to address intracranial metastases, the 91-year-old president experienced complete regression of his tumour. Now, at the age of 96, he remains cancer-free (Cancer Research Institute, 2020).

Beyond melanoma, immune checkpoint inhibitors have demonstrated activity against a wide variety of solid tumours including non-small-cell lung cancer, which by itself is the single most important cause of cancer-related death. Pembrolizumab, as one example, is now indicated for cancer treatment in 30 different settings across 16 different tumour types (FDA, 2020a). The availability of improved therapies will meaningfully accelerate the already impressive decline in cancer death rates that has resulted from smoking cessation. Indeed, in 2021 the American Cancer Society announced that from 2017 to 2018, the most recent year for which data were available, the cancer death rate declined by 2.4% (American Cancer Society, 2021).

Much of the reduction in cancer death rates relates to lung cancer, which is the leading cause of cancer death worldwide. Non-small-cell lung cancer accounts for about 85% of all lung cancer cases, and until recently most such patients had a very poor prognosis. Today, the hazard ratio for overall survival in locally advanced or metastatic non-small-cell lung cancer treated with checkpoint inhibitors, often administered with chemotherapy, as compared with chemotherapy alone has been estimated at 0.72 across a meta-analysis involving 12 different studies (Liang et al., 2020). Hence, this treatment alone can be expected to drive a nearly 30% reduction in death rates for the most lethal human malignancy.

In sum, it is plain that new treatments and vaccines can greatly extend life expectancy, especially for those who have reached middle age. A recent and especially important example has been the successful development of drugs and vaccines for the control of COVID-19 pandemic, which in 2020 was directly responsible for more than 300,000 deaths just in the USA—largely in the elderly population—and had devastating consequences on life expectancy, especially in Black and Latino populations (Andrasfay and Goldman, 2021). Through a focused effort involving hundreds of corporations, nongovernmental organizations, and government agencies, multiple efficacious COVID-19 vaccines were developed in less than a year (Baden et al., 2020; Polack et al., 2020), and in the same period several anti-COVID-19 agents were also shown to improve survival in severely affected patients (Beigel et al., 2020).

These rapid successes are the exception rather than the rule in drug development. While effective drugs and vaccines are critical components of a healthcare infrastructure, the pharmaceutical industry itself is struggling in its efforts to discover and develop new therapies.

Drug discovery is an increasingly difficult undertaking

The development of drugs and vaccines is a complicated endeavour, requiring deep understanding of human biology at the molecular level. Advances in organic chemistry and genetic engineering have produced new agents characterized by remarkable selectivity. It is this selectivity that creates a therapeutic window, such that the benefits resulting from the treatment are realized at levels of exposure far below those associated with adverse effects.

As noted above, new drug discovery can have a very positive impact on conventional measures of health. In the field of cancer, for example, one can evaluate the years of potential life lost before age 85 as a function of new drugs introduced. One such analysis, conducted in New Zealand for the years 1998 to 2017, demonstrated that the introduction of new cancer therapies provided unquestionable public health benefits with a favourable cost/benefit ratio (Lichtenberg, 2021). That the analysis was conducted before the most recent wave of immunomodulatory therapies which, as I have noted, are especially active in highly lethal tumour settings, suggests that societies are assessing the value of these innovations correctly.

This said, the productivity of the pharmaceutical industry, measured as the yield of new drugs per billion dollars of research expenditure, has declined exponentially for most of the last century (Manos, 2009). Through at least the midpoint of the last decade there has been no evidence of moderation in the cost of biopharmaceutical innovation. To this point, a 2016 analysis of research and development costs in the biopharmaceutical industry conducted by the Tufts Center for the Study of Drug Development estimated that each new drug, on average, requires an investment of approximately $2.6 billion (both out-of-pocket and capitalized expenditures in 2013 dollars) to gain regulatory approval (DiMai et al., 2016). This number reflects the actual costs incurred to discover and test new pharmaceutical compositions in preclinical and clinical settings, as well as the cost associated with developing the much larger set of compounds that failed to achieve the desired therapeutic effect (the 'cost of failure'). Adjustment for post-approval costs and correction for inflation places the research and development investment at more than $3 billion per new drug, an astonishing number. Moreover, despite great improvements in our ability to measure the effects of drug candidates on disease outcomes, the fraction of new therapeutic candidates that succeed in delivering favourable benefits has remained constant, or perhaps has even declined. It is this enormous cost burden that undermines the viability of biopharmaceutical companies generally (Schuhmacher et al., 2016). Simply put, most companies committed to discovering and developing new medicines appear doomed by the low productivity of their research organizations.

An important exception to this general rule lies in the area of rare diseases, notably in the area of diseases that result from heritable defects in individual genes. In many cases it is possible to replace a defective gene product with an exogenously produced protein that provides the essential missing function. For example, haemophilia A or B, diseases of coagulation that result from crippling mutations in the factor VIII or factor IX genes, respectively, of the coagulation cascade (Kulkarni and Soucie, 2011), are now treated with exogenously administered proteins produced in bioreactors (e.g. Valentino et al., 2014). There are more than 2,000 diseases with defined phenotypes that are characterized by Mendelian inheritance and hence attributable to single-gene defects (National Library of Medicine, 2020). Each of these diseases affect relatively few individuals, and even in aggregate account for a small fraction of human morbidity. Nevertheless, under circumstances where the relationship between the gene product and the disease manifestations is well understood, it stands to reason that early replacement of the defective gene product—or more recently the gene itself—will have beneficial effects. The probability of success for drug discovery efforts is improved immeasurably under these circumstances, and the clinical studies necessary to support commercialization are also, understandably, much abbreviated.

These considerations have stimulated a change in research activity for pharmaceutical companies. Considering only those novel agents approved by the US Food and Drug Administration (FDA, 2020b), of 44 new molecular entities, excluding four approvals for drugs employed in diagnosing disease, fully one-quarter were developed for exceedingly rare genetic diseases ranging from Duchenne muscular dystrophy—golodirsen (Young, 2020) to acute hepatic porphyria—givosiran (Balwani et al., 2020).

Although the development efforts necessary to produce these drugs require substantially less investment than those required for agents intended to treat more common diseases, the per-patient cost for an approved medicine is typically quite high. Thus, the gene therapy Zolgensma™ (onasemnogene abeparvovec), approved by the FDA in 2019 for the treatment of spinal muscular atrophy (Stevens et al., 2020), was priced at US$2.1 million (Thomas, 2019), the most expensive drug in history. This extraordinary price may compare favorably with that of Spinraza™, approved for the same condition in 2016, which reportedly costs US$750,000 in the first year, and US$350,000 for each year of life thereafter (Picchi, 2016). The long-term outcomes in children diagnosed with spinal muscular atrophy, representing about one in 10,000 live births in the USA (Sugarman et al., 2011), will become known over time, but the economic burden of administering these treatments to all affected individuals, even when weighed against the prior cost of caring for sick children who otherwise had no treatment options, exceeds what most healthcare systems can support.

Why preventive medicine cannot ensure increased life expectancy

Leaving aside pricing, the fact that the biopharmaceutical industry now devotes a substantial fraction of its research efforts towards the treatment of rare genetic diseases should raise concerns regarding the ability to improve outcomes for patients suffering from those diseases that are major causes of morbidity and mortality, especially cardiovascular disease, cancer, dementia, and psychiatric disease.

The plain fact, of course, is that most people are healthy most of the time. It has long been recognized that patients generally enter the healthcare system only when they have developed relatively advanced disease, at a time when preventive measures, e.g. smoking cessation and dietary interventions, will almost certainly not be sufficient to alter the course of illness. It stands to reason that if we wish to secure improved life expectancy we will need to find ways to identify disease propensities early, and to modify those factors that contribute to the progression of illness. Therein lies the envisioned promise of P4 medicine: genome sequencing, perhaps from birth, provides an understanding of individual disease propensity, and lifestyle modifications thereafter improve the likelihood of leading a disease-free life for the longest possible time.

This promise notwithstanding, no such system, however thoroughly implemented, will free human populations from suffering and disease. While continued effort to characterize inherited predisposition to disease is necessary, and further studies designed to illuminate salubrious lifestyle modifications are essential, it is nevertheless the case that stochastic processes often underlie ill health. These include the effects of extrinsic inputs, e.g. exposure to infectious agents, as well as the consequences of variations in human development that result from random deviations in normal cellular behaviour.

Consider, for example, the development of malignant disease in an otherwise healthy individual. Cancer ranks second among the leading causes of death in economically advanced countries. Malignant transformation of cells occurs because of the accumulation of somatic mutations which, in aggregate, confer upon the cell malignant properties, e.g. growth factor independence, invasiveness, metastatic potential, and genomic instability. In many cases, the mutations that give rise to cancer are associated with exposure to environmental carcinogens, like those found in tobacco smoke. However, the frequency of replication errors during normal cell division is sufficiently high to ensure that replicating cells will occasionally, during the life of the individual, give rise to cells that embark upon a stepwise progression to malignant behaviour (Tomasetti et al., 2017). Just as variations in physical body plan are a normal consequence of physiological development, disadvantageous variations in normal physiology will accrue as a consequence of normal development: human beings are not products of a rigidly imposed manufacturing scheme.

Even more frequently, some interaction between external influences such as smoking and internal variability (e.g. the inheritance of predisposing mutations) will lead to the accelerated acquisition of a disease phenotype. Involvement of both genetic and environmental factors in the pathogenesis of cardiovascular disease has been well-documented (Kathiresan and Srivastava, 2012). Resistance to cardiovascular disease in such individuals depends in part upon the adoption of a healthy lifestyle, but will inevitably also require therapies that reduce serum LDL-cholesterol levels (Goldman et al., 2019), as well as new therapies that act upon targets that we have not yet identified.

For these reasons, it is plain that assembling more data on disease propensity is especially valuable when disease treatments improve as a result. Absent the availability of new therapies, it will not be possible to eliminate the grievous illnesses that most reduce human life expectancy.

What we can expect from fundamental research

There is broad evidence that environmental exposures as well as genetic variation both contribute to the burden of neurodegenerative and psychiatric diseases. Whereas inheritance plays some role in the propensity to develop these diseases, the genome is rarely deterministic. Consider the case of Parkinson's disease, a devastating neurodegenerative illness that causes severe rigidity and cognitive impairment. Studies of identical twins reveal that concordance for Parkinson's disease is quite modest—less than 20% in most series (Eriksen et al., 2005). Indeed, in this instance the nearly identical concordance observed in dizygotic twin pairs, as compared with monozygotic (identical) twins, suggests that shared environmental effects—perhaps experienced quite early in life—contribute to the development of Parkinson's disease. Hence if we wish to treat Parkinson's disease effectively, we must have some insight into the pathologic processes that cause degeneration of dopaminergic neurons in the central nervous system, the proximate cause of this devastating disorder (Eriksen et al., 2005). This in turn will permit hypotheses to be tested regarding means of interdicting established disease, and will ultimately permit the evolution of preventive measures that reduce the likelihood of development of Parkinson's disease later in life.

Irrespective of how successfully we implement cradle-to-grave data collection and analysis, for the foreseeable future it will not be possible to identify many diseases at an early point. Even when these technologically advanced diagnoses are achievable, lifestyle modifications will rarely by themselves arrest disease progression. Instead, the control of most severe illnesses will require the development of novel therapeutic interventions. The challenge in confronting this problem results from the fact that we have little insight into normal human physiology, much less the pathologic processes, operating at cellular and molecular levels, that contribute to disease. In this sense, it is something of a miracle whenever an experimental therapeutic achieves a favourable benefit/risk profile. This, too, helps explain why pharmaceutical companies have directed their resources towards rare, heritable diseases. In these monogenic illnesses one can usually draw a clear line between the molecular defect, defined at the level of the responsible gene itself, and the pathology that eventually emerges.

As DNA sequence determination becomes even less expensive, and as millions of individual genome sequences are correlated with clinical outcomes over decades, there is every reason to believe that the molecular understanding of disease pathogenesis will increase enormously. With sufficient investment in resources and time, the physiologic bases of most diseases will become known, and therapies will be tailored to alleviate the impact of these molecular pathologies. This fundamental research, still largely supported by public funds and by philanthropic donations to non-governmental agencies, remains the hope for all those who wish to reduce the suffering of human populations. A healthy pharmaceutical sector as well, drafting behind the fundamental research supported by governments, provides the disciplined infrastructure necessary to produce advanced therapeutics.

Implications of research investment for the health of human societies

Technological advancement is improving the quality of medical practice. First, more precise descriptions of diseases, aided by better measurement tools, are permitting critically important distinctions to be made among otherwise indistinguishable diseases. Consider, for example, the heterogeneity of malignant lymphomas. So-called non-Hodgkin lymphomas represent the sixth most common cancer among men and women, with an incidence of about 19.6 cases per 100,000 individuals per year, which has been stable for the last quarter-century (National Cancer Institute, 2020). The nosology of these lymphomas has, however, changed markedly during this period. First, it became possible to identify cellular subtypes of non-Hodgkin lymphomas with more homogeneous characteristics. Thus we now speak of diffuse large B-cell lymphoma as the most common type of non-Hodgkin lymphoma (Wright et al., 2020). Within this subtype, genetic analysis permits much finer discrimination of cancers that respond to therapy in a similar way: tumours that manifest certain types of genetic abnormalities have very different prognoses and must be treated very differently. In the absence of therapy, few patients afflicted with diffuse large B-cell lymphoma will survive much longer than two years (Rovira et al., 2014). However, with aggressive combination chemotherapy tailored to the genetic make-up of the tumour, this disease is now largely curable, with 5-year relative survival numbers having improved from 45% in 1975 to nearly 80% in 2018 (Pulte et al., 2020). Hence no competent oncologist in those countries with advanced medical practice would consider treating this disease without an assessment of the genetic make-up of the tumour cells.

But this type of molecular medicine has been slow to penetrate low-income countries. The current cost for sequencing an entire human genome has fallen below US$1,000 in the USA and Europe, with predictions that this cost will reach US$100 in just a few years (National Genome Institute, 2020). But even US$100 far exceeds the annual per-capita expenditure on healthcare for much of the world's population; in 2016 more than 10% of the world's population lived with annual per-capita healthcare spending of $40, and modelling suggests that the fraction of the global population confined to these low spending tiers will increase over the next three decades (Chang et al., 2019). Without a concentrated effort, we can expect that the technological advances that permit personalized medicine will further divide the world's countries into those whose citizens realize greater health benefits, and those for whom highly individualized medical practice is simply unaffordable. This is true for practices that align data-gathering with lifestyle management, and of course for those circumstances where novel, life-extending therapies, typically priced for high-income markets, become available.

Beyond these differences, the development of personalized medicine will alter how we think about disease itself. Consider, as an example, the potential evolution of psychiatric practice that could derive from the application of artificial neural networks to psychiatric diagnosis and treatment. These approaches, taking advantage of data-rich information from individual responses to images and to game-playing, could permit accurate identification of those at near-term risk for suicide, or for displaying violent behaviour (Durstewitz et al., 2019). Treatment algorithms for these individuals would likely change as well. Variations in patterns of human thinking are complex, and in many ways too challenging to be addressed by human intelligence. Machine-learning approaches, however, may permit recognition of common patterns in formal thought disorders (Lei et al., 2019).

An objective approach to psychiatric disease taking advantage of artificial neural networks might be expected to alter the way in which we regard free will, and hence could affect long-held perspectives on imprisonment and re-habilitation. Machine-learning algorithms designed to assist in the diagnosis and treatment of psychiatric disease might also prove to be more effective tools for monitoring human performance. This sort of rumination impinges powerfully upon what it means to be human, and what we can hope for as individuals and as human societies. In simple terms, rigorous measurement of thousands of physiologic parameters in millions of individuals will inev-itably provide an extremely detailed picture of individual capabilities, both physical and intellectual. It remains to be seen whether such data sets can be harnessed for the benefit of any individual, much less for society as a whole.

Massive improvements in computational power, at ever lower cost, are transforming our world. Data obtained from ordinary mobile phones, al-ready in widespread use throughout the world, can permit assessment of changes in physiologic state and mood, essentially in real time (Haines-Demont et al., 2020). While currently used to provide applications for per-sonal improvement (monitoring exercise performance or sleep quality), we can expect that quite simple adjustments will successfully predict the risk of suicidal behaviour in depressed individuals with a prior history of suicide attempts, or violent behaviour in those predisposed to sociopathy, at once providing a propensity score for adverse behaviour and the precise location of the individual at risk. Since violence of this type contributes importantly to years of healthy life lost, as illustrated in a careful study performed using data from 2006 to 2013 in Rhode Island (Jiang et al., 2016), will governments not feel compelled to act on this information? What, then, constrains the intro-duction of machine supervision into the routine activities of daily living?

Finally, it must be pointed out that demographic changes that attend im-proved management of disease are transforming the world. For the first time in human history, the average age of the human population, on every continent, is increasing meaningfully (UN Department of Economic and Social Affairs, 2019). As an example, the US Census Department (2018) has announced that by 2034 the population of those aged above 65 will exceed those from birth to age 18 for the first time in US history. It is plain that older people will increasingly dominate human societies, not just in the economically advantaged countries, but also in low-income countries. Since serious illness rates increase markedly in those above 60 years of age, governments already challenged by high healthcare expenditures will find it necessary to ration the administration of lifesaving interventions even more strictly. The effort to balance the needs of the elderly with the productive output of young people will occupy substantial effort on the part of human societies for the foreseeable future. This is in part a consequence of the successful application of personalized medicine, which also represents the principal hope for future improvement in the overall quality of human life through the development of safer, and more effective, pharmaceuticals.

References

ABIM, ACP–ASIM, and European Federation of Internal Medicine. 2002. Medical professionalism in the new millennium: a physicians' charter. *Ann. Intern. Med.* 136, 243–6.

ABIM Foundation, ACP-ASIM Foundation, and European Federation of Internal Medicine. 2002. Medical professionalism in the new millennium: a physician's charter. *Acta Clin. Belg.* 57, 169–71.

Acheoah, J. E. and Hamzah, A. 2015. Style in Christian and Islamic sermons: a linguistic analysis. *Am. Res. J. Engl. Lit.* 1, 23–31. See utterance 17.

Ackerman, M. J. 2015. Genetic purgatory and the cardiac channelopathies: exposing the variants of uncertain/unknown significance issue. *Heart Rhythm* 12, 2325–31.

Adams, W. Y. 1998. *The Philosophical Roots and Anthropology.* Stanford: CSLI Publications.

Adashi, E. Y. and Cohen, G. 2020. Heritbale human genome editing: the international commission report. *JAMA* 324, 1941–2.

Adhikary, P., Keen, S. and Van Teijlingen, E. 2011. Health issues among Nepalese migrant workers in the Middle East. *Health Sci. J.* 5, 169–75.

Agriantonis, D. J., Hall, L. and Wilson, M. A. 2008. Pitfalls of I-131 whole body scan interpretation: bronchogenic cyst and mucinous cystadenoma. *Clin. Nucl. Med.* 33, 325–7.

Ahmed, A. and Suleman, M. 2018. Islamic perspectives on the genome and the human person: why the soul matters. In Ghaly, M. (ed.), *Islamic Ethics and the Genome Question*, pp. 139–68. Leiden: Brill.

Ahn, H. S., Kim, J. J. and Welch, G. 2014. Korea's thyroid cancer "epidemic"—screening and over diagnosis. *N. Engl. J. Med.* 371, 765–7.

Al Aqeel, A. I. 2007. Islamic ethical framework for research into and prevention of genetic diseases. *Nature Genet.* 39, 1293–8.

Alan, R. 2013. Biological races in humans. *Stud. Hist. Phil. Sci.* 44C, 262–71.

Allport, G. 1965. *Pattern and Growth in Personality.* New York: Holt.

Al-Shahri, M. Z., and Al-Khenaizan, A. 2005. Palliative care for Muslim patients. *J. Support Oncol.* 3, 432–6.

Alter, J. S. 1999. Heaps of health, metaphysical fitness: Aryuveda and the ontology of good health in medical anthropology. *Curr. Anthropol.* 40, S43–S66.

Altman, R. 2015. Foreword: Biology's Love Affair with the Genome. In Richardson, S. and Stevens, H. (eds), *Postgenomics: Perspectives on Biology After the Genome*, pp. vii–ix. Durham, NC/London, Duke University Press.

Alyass, A., Turcotte, M. and Meyre, D. 2015. From big data analysis to personalized medicine for all: challenges and opportunities. *BMC Med. Genom.* 8, 33.

American Cancer Society. 2021. Facts & figures 2021 reports another record-breaking 1-year drop in cancer deaths. Available at: https://www.cancer.org/latest-news/facts-and-figures-2021.html

American Psychiatric Association. 2013. Diagnostic and Statistical Manual of Mental Disorders, 5th edn. Washington: American Psychiatric Publishing.

Anand, G. 2009. The Cure: How a Father Raised $100 Million—and Bucked the Medical Establishment—in a Quest to Save His Children. New York: Harper Collins.

Andrasfay, T. and Goldman, N. 2021. Reductions in 2020 US life expectancy due to COVID-19 and the disproportionate impact on the Black and Latino populations. Proc. Natl Acad. Sci. USA 118(5), e2014746118.

Angrist, M. 2010. Here is a Human Being: At the Darn of Personal Genomics. New York: HarperCollins.

Angrist, M., Barrangou, R., Baylis, F., et al. 2020. Reactions to the National Academies/Royal Society Report on Heritable Human Genome Editing. CRISPR J. 3, 333–49.

Ankeny, R. A. and Leonelli, S. 2013. Valuing data in postgenomic biology: How data donation and curation practices challenge the scientific publication system. In: Solomon, S. S. and Stevens, H. (eds), Postgenomics: Perspectives on Biology after the Genome, pp. 126–49. Durham: Duke University Press.

Annas, G. 2000. Rules for research on human genetic variation—lessons from Iceland. N. Engl. J. Med. 342, 1830–3.

Anonymous 2019a. Gene therapy's next installment. Nature Biotechnol. 37, 697.

Anonymous 2019b. Tackle sickle cell economics. Nature 576, 7–8.

Apweiler, R., Beissbarth, T., Berthold, M. R., et al. 2018. Whither systems medicine? Exp. Mol. Med. 50, e453.

Arendt, H. 2006. Between Past and Future. New York: Pengin Books.

Arendt, H. 1998. The Human Condition. Chicago: University of Chicago.

Arendt, H. 2006. Eichmann in Jerusalem: A Report on the Banality of Evil. New York: Penguin.

Arendt, H. 1978. The Life of the Mind. New York, NY: Harcourt, Inc.

Arendt, H. 1994. On the Nature of Totalitarianism: An Essay in Understanding. In Essays in Understanding, 1930–1954. New York, New York: Schocken Books.

Arno, A. and Thomas, S. 2016. The efficacy of nudge theory strategies in influencing adult dietary behavior: a systematic review and meta-analysis. BMC Public Health 16, 676.

Arnold, A. P., Chen X. and Itoh Y. 2012. What a difference an X or Y makes: sex chromosomes, gene dose, and epigenetics in sexual differentiation. Handb Exp Pharmacol. 214, 67–88.

Arnold, A. P. and Lusis, A. 2012. Understanding the Sexome: Measuring and Reporting Sex Differences in Gene Systems. Endocrinology 153, 2551–5.

Ashley, E. A. 2015. The precision medicine initiative: a new national effort. JAMA 313, 2119–20.

Augspurger, E. E., Rana, M. and Yigit, M. V. 2018. Chemical and biological sensing using hybridization chain reaction. ACS Sensors 3, 878–902.

Austin, C. P. 2018. Translating translation. Nat. Rev. Drug Discov. 17, 455.

Badano, G. 2016. Still special, despite everything: a liberal defense of the value of healthcare in the face of the social determinants of health. *Social Theory Pract.* 42, 183–204.

Baden, L. R., El Sahly, H. M., Essink, B., et al. 2020. Efficacy and safety of the mRNA-1273 EARS-CoV-2 vaccine. *N. Engl. J. Med.* 384, 403–16.

Baier, A. C. 1986. Trust and antitrust. *Ethics* 96, 231–60.

Baltimore, D., Berg, P., Botchan, M., et al. 2015. A prudent path forward for genomic engineering and germline gene modification. *Science* 348, 36–8.

Balwani, M., Sardh, E., Ventura, P., et al. 2020. Phase 3 trial of RNAi therapeutic givosiran for acute intermittent porphyria. *N. Engl. J. Med.* 382, 2289–301.

Banu az-Zubair, M. K. 2007. Who is a parent? Parenthood in Islamic ethics. *J Med. Ethics* 33, 605–9.

Barad, K. 2007. *Meeting the Universe Halfway: Quantum Physics and the Entanglement of Matter and Meaing.* Durham and London: Duke University Press.

Barbour, D. L. 2019. Precision medicine and the cursed dimensions. *NPJ Digital Med.* 2, 4.

Bardiga, I. 2008. Reading the universal book of nature: the *Accademia dei Lincei* in Rome (1603–1630). In van Dixhoorn, A. and Speakman Sutch, S. (eds), *The Read of The Republic of Letters: Literary and Learned Societies in Late Medieval and Early Modern Europe,* Vol. 2, pp. 353–388. New York: Brill.

Barilan, Y. M. 2003. One or two: an examination of the recent case of the conjoined twins from Malta. *J. Med. Phil.* 28, 27–44.

Barilan, Y. M. 2005. The debate on cloning: some contributions from the Jewish tradition. In: Roetz, H. and Frey, C. (eds), Cloning: a multicultural perspective, pp. 311–40. Berlin/London: Rodopi.

Barilan, Y. M. 2012. *Human Dignity, Human Rights and Responsibility: The New Language of Bioethics and Biolaw.* Cambridge, MA: MIT Press.

Barilan, Y. M. 2014. *Jewish Bioethics: Rabbinic Law and Theology in Their Social and Historical Contexts.* Cambridge: Cambridge University Press.

Barker, D. J., Gluckman, P. D., Godfrey, K. M., et al. 1993. Fetal nutrition and cardiovascular disease in adult life. Lancet 341, 938–41.

Barnouw, V. 1985. *Culture and Personality,* 4th edn. Homewood: Dorsey Press.

Barros, J. R. 2016. UCSF Plans to Shutter Clinic Serving Minority Youth. *Mission Local,* March 15; available at: https://missionlocal.org/2016/03/ucsf-plans-to-shutter-clinic-serving-minority-youth/

Bastable, S. B., Sopczyk, D., Gramet, P., Jacobs, K. 2011. Health Professionals as Educator: Principles of Teaching and Learning. Sudbury, MA: Jones & Bartlett Learning, LLC.

Bayer, R. and S. Galea. 2015. Public health in the precision-medicine era. *N. Engl. J. Med.* 373, 499–501.

Beckmann, J. S. and Lew, D. 2016. Reconciling evidence-based medicine and precision medicine in the era of big data: challenges and opportunities. *Genome Med.* 8, 134.

Beigel, J. H., Tomashek, K. M., Dodd, L. E., et al. 2020. Remdesivir for the treatment of COVID-19—Final Report. *N. Engl. J. Med.* 383, 1813–26.

Bellman, R. E. 1957. *Dynamic Programming*. Princeton, NJ: Princeton University Press.

Bendich, A. 1966. Privacy, poverty and the constitution. *Calif. Law Rev.* 54, 407–42.

Benjamin, R. M. 2012. Medication adherence: helping patients take their medicine as directed. *Public Health Rep.* 127, 2–3.

Bera, A., Randhavane, T. and Maocha, D. 2017. Aggressive, tense or shy? Identifying personality traints from crowd video. *Proc. Twenty-Sixth Int. Joint Conf. on Artificial Intelligence*, Melbourne, Australia, 19–25 August 2017, pp. 112–18.

Berg, P. 2008. Asilomar 1975: DNA modification secured. *Nature* 455, 290.

Bewley, S. 2006. Genetic profiling of newborns: ethical and social issues. *Nature Rev.* 7, 67–71.

Bhatt, A. 2019. New clinical trial rules: academic trials and tribulations. *Perspect. Clin. Res.* 10, 103.

Biesecker, L. G. Burke, W., Kohane, I., et al. 2012. Next generation sequencing in the clinic: are we ready? *Nat. Rev. Genet.* 13, 818–24.

Biesecker, L. G. and Green, R. C. 2014. Diagnostic clinical genome and exome sequencing. *N. Engl. J. Med.* 370, 2418–25.

Black, J. 2001. Decentring regulation: understanding the role of regulation and self-regulation in a "post-regulatory" world. *Curr. Legal Problems* 54, 103–46.

Black, J. 2014. Learning from regulatory disasters. *Policy Q.* 10, 1–11.

Blanchard, A. P., Kaiser, R. J. and Hood, L. 1996. High-density oligonucleotide arrays. *Biosens. Bioelectron.* 11, 687–90.

Blasco-Fontecilla, H. 2014. Medicalization, wish-fulfilling medicine, and disease mongering: toward a brave new world? *Rev. Clin. Esp. (Barc.)* 214, 104–7.

Blasimme, A. and Vayena, E. 2016. Becoming partners, retaining autonomy: ethical considerations on the development of precision medicine. *BMC Med Ethics* 17, 67.

Blumenthal, G. M. and Pazdur, R. 2018. Approvals in 2017: gene therapies and site-agnostic indications. *Nature Rev. Clin. Oncol.* 15, 127–28.

Boethius. 1847. In Porphyrium Dialogi a Victorino Translati. In: Migne, J-P. (ed.), Patrologiae Cursus Completus, Series Prima, vol. 64. Paris.

Boggio, A., Romano, C. and Almqvist J. (eds) 2020. *Human Germline Modification and the Right to Science: A Comparative Study of National Laws and Policies.* Cambridge: Cambridge University Press.

Bolton, J. W. 2014. Case formulation after Engel—The 4P model: a philosophical case conference. *Philosophy Psychiatry & Psychology* 21, 179–89.

Bok, P. K. 1988. *Rethinking Psychological Anthropology: Continuity and Change in the Study of Human Action.* New York: Freeman.

Bonham, V. L., Callier, S. L. and Royal, C. D. 2016. Will precision medicine move us beyond race? *N. Engl. J. Med.* 374, 2003–5.

Boniolo, G. 2017a. Molecular medicine: the clinical method enters the lab. In: Boniolo, G. and Nathan, M. (eds), Philosophy of Molecular Medicine: Foundational Issues in Research and Practice, pp. 15–34. New York: Routledge.

Boniolo, G. 2017b. Patchwork narratives for tumour heterogeneity. In: Leitgeb, H., Niiniluoto, I., Sober, E. and Seppälä, P. (eds), Proceedings of the 15th International Congress on Logic, Methodology and Philosophy of Science, pp. 311–24. London: College Publications.

Boniolo, G. and Campaner, R. 2019. Causal reasoning and clinical practice: challenges from molecular biology. *Topoi* 38, 423–35.

Boniolo, G. and Campaner, R. 2020. Life sciences for philosophers and philosophy for life scientists. What should we teach? *Biol. Theory* 15, 1–11.

Boniolo, G. and Lorusso, L. 2008. Clustering humans: on biological boundaries studies. *Stud. Hist. Phil. Biol. Biomed. Sci.* 39, 63–70.

Boniolo, G. and Sanchini, V. (eds) 2016. Ethical Counselling and Medical Decision-Making in the Era of Personalized Medicine, Heidelberg: Springer.

Boniolo, G. and Teira, D. 2016. The centrality of probability. In: Boniolo, G. and Sanchini, V. (eds), Ethical Counselling and Medical Decision-Making in the Era of Personalized Medicine, pp. 49–62. Heidelberg: Springer.

Boorse, Ch. 1997. A rebuttal on health. In: Humber, J. M. and Almeder, R. F. (eds), What Is Disease?, pp. 1–134. New York: Springer.

Booth, N., Robinson, P. and Kohannejad, J. 2004. Identification of high-quality consultation practice in primary care: the effects of computer use on doctor–patient rapport. *Inform. Prim. Care* 12, 75–83.

Bossen, C., Pine, K. H., Cabitza, F., et al. 2019. Data work in healthcare: an introduction. *Health Informatics J.* 25, 465–74.

Boston Women's Health Book Collective. 2011 (1st edn: 1973). *Our Bodies, Ourselves.* New York: Simon & Schuster.

Botkin, J. R., Lewis, M. H., Watson, M. S., et al. 2014. Parental permission for pilot newborn screening research: guidelines from the NBSTRN. *Pediatrics* 133, e410–e417.

Bouk, D. B. 2015. *How Our Days Became Numbered.* Chicago: University of Chicago Press.

Bousquet, J., Anto, J. M., Sterk, P. J., et al. 2011. Systems medicine and integrated care to combat chronic noncommunicable diseases. *Genome Med.* 3, 43.

Brattain, M. 2007. Race, Racism, and Antiracism: Unesco and the Politics of Presenting Science to the Postwar Public. *Am Hist Rev.* 112(5), 1386–413.

Braun, B. 2007. Biopolitics and the molecularization of life. *Cult. Geogr.* 14, 6–28.

Bredström, A. 2019. Culture and context in mental health diagnosing: scrutinizing the DSM-5 revision. *Med. Humanit.* 40, 347–63.

Brenner, S. 2010. Sequences and consequences. *Phil. Trans. R. Soc.* 365B, 207–12.

Brisson, M., Kim, J. J., Canfell, K., et al. 2020. Impact of HPV vaccination and cervical cancer screening on cervical cancer elimination: a comparative modelling analysis on 79 low-income and lower-middle-income countries. *Lancet* 395, 575–90.

Broadbent, A. 2017. Philosophy of epidemiology. In Marcum, J. A. (ed.), The Bloomsbury Companion to Contemporary Philosophy of Medicine, ch. 4, pp. 93–112. London: Bloomsbury Academic.

Brodersen, J., Schwartz, L. M. and Woloshin, S. 2014. Overdiagnosis: how cancer screening can turn indolent pathology into illness. *APMIS* 122, 683–9.

Brodersen, J., Schwartz, L. M., Heneghan, C., et al. 2018. Overdiagnosis: what it is and what it isn't. *BMJ Evid. Based Med.* 23, 1–3.

Bruce, G. and Critchley, C. 2008–2017. *Swinburne National Technology and Society Monitor* Melbourne, Australia: Swinburne University of Technology. Available at: https://researchbank.swinburne.edu.au/items/4d553f11-b48c-4032-9505-cd0da8eeef9d/1/

Brundell, B. 2001. Catholic Church politics and evolutionary theory. *BJHS* 34, 81–95.

Brungs, R. A. 1995. Hyrbrid, Genes and Patents. *ITEST Bulletin* 26(3), 3–16.

Brusa, M. and Barilan, Y. M. 2017. Newborn screening on the cusp of genetic screening: from solidarity in public health to personal counselling. In: Petermann, H. I., Harper, P. S. and Doetz, S. (eds), *History of Human Genetics, Aspects of Its Development and Global Perspectives*, pp. 503–22. New York: Springer.

Budrionis, A. and Bellika, J. G. 2016. The learning healthcare system: where are we now? A systematic review. *J. Biomed. Informatics* 64, 87–92.

Buttorff, C., Ruder, T. and Bauman, M. 2017. Multiple chronic conditions in the United States. Santa Monica, CA: Rand Corp.

Cabibbo, N. 2004. The meaning of the Pontifical Academy of Sciences. *The Pontifical Academy of Science Acta 17*, 115–20.

Cajka, P. 2021. This May Help Us and the Whole Cause of Due Process: Father Joseph Byron, the Washington Nineteen, and American Catholic Priests as Sixties Rebels. *The Sixties: A Journal of History, Politics and Culture* 14:1, 74–96.

Calnan, M. and Rowe, R. 2008. *Trust matters in health care*. Buckingham: Open University Press.

Campos-Outcalt, D. 2008. Transforming health care through prospective medicine: a broader perspective is needed. *Acad. Med.* 83, 705.

Cancer Research Institute. 2020. *Jimmy Carter's Cancer Immunotherapy Story*. Available at: https://www.cancerresearch.org/join-the-cause/cancer-immunotherapy-month/30-facts/20

Carrington, C. H. 2006. Clinical depression in African American women: diagnoses, treatment, and research. *J. Clin. Psychol.* 62, 779–91.

Cassell, E. J. 1993. The sorcerer's broom. *Hastings Center Rep.* 23, 32–9.

Caulfield, T. 2015. Genetics and personalized medicine—where's the revolution? July 23. Available at: http://blogs.bmj.com/bmj/2015/07/23/timothy-caulfield-genetics-and-personalized-medicine-wheres-the-revolution/

Caulfield, T. and Murdoch, B. 2017. Genes, cells, and biobanks: yes, there's still a consent problem. *PLoS Biology* 15, e2002654.

Ceyhan-Birsoy, O., Murry, J. B., Machini, K., et al. 2019. Interpretation of genomic sequencing results in health and ill newborns: results from the BabySeq project. *Am. J. Hum. Genet.* 104, 76–93.

Chadwick, R. and Berg, K. 2001. Solidarity and equity: new ethical frameworks for genetic databases. *Nature Rev. Genet.* 2, 318–21.

Chakravarti, A. 2011. Genomics is not enough. *Science* 334, 15.

Champer, J., Buchman, A. and Akbari, O. S. 2016. Cheating evolution: engineering gene drives to manipulate the fate of wild populations. *Nature Rev. Genet.* 17, 146–59.

Chang, A. Y., Cowling, K., Micah, A. E. et al. 2019. Past, present, and future of global health financing: a review of development assistance, government, out-of-pocket, and other private spending on health for 195 countries, 1995–2050. *Lancet* 393, 2233–60.

Charoenngam, N. and Holick, M. F. 2020. Immunologic effects of vitamin D on human health and disease. *Nutrients* 12(7), 2097. doi: 10.3390/nu12072097. PMID: 32679784; PMCID: PMC7400911.

Chen, R., Mias, G. I., Li-Pook-Than, J., et al. 2012. Personal omics profiling reveals dynamic molecular and medical phenotypes. *Cell* 148, 1293–307.

Cherlo, Y. 2002. Halakha Hanevuit, Akdamot, vol. 12. Jerusalem. [Hebrew]

Childress, J. F. 1997. *Practical Reasoning in Bioethics*. Bloomington, IN: Indiana University Press.

Chisholm, A. M., Montgomery, C., Parkin, S. and Locock, L. 2020. Wild data: how frontline hospital staff make sense of patients' experiences. *Sociol. Health Illness* 42, 1424–40.

Chneiweiss, H., Hirsch, F. and Montoliu, L. 2018. Fostering responsible research with genome editing technologies: a European perspective. *Transgenic Res.* 26, 709–713.

Chowkwanyun, M., Bayer, R. and Galea, S. 2018. "Precision" public health—between novelty and hype. *N. Engl. J. Med.* 379, 1398–400.

Christianson, A., Gribaldo, L., Harris, H., et al. 2013. Genetic testing in emerging economies (GenTEE), scientific and policy report by the Joint Research Centre of the European Commission 37, 62. doi: 10.2788/26690

Chrystal, A. and Mizen, P. 2003. Goodhart's Law: its origins, meaning and implications for monetary policy. In P. Mizen (ed.), *Central Banking, Monetary Theory and Practice*, pp. 221–43. Cheltenham: Edward Elgar.

Ciriello Pothier, K. 2017. Personalizing Precision Medicine. New York: Wiley.

Clark, P. A., Friedman, J. R., Crosson, D. W. and Fadus, M. 2011. Concierge medicine: medical, legal and ethical perspectives. *Internet J. Law, Healthcare Ethics* 7, 1528–8250.

Claw, K. G., Matthew, Z., Anderson, R. L., et al. 2018. A Framework for Enhancing Ethical Genomic Research with Indigenous Communities. *Nat Commun* 9, 2957.

Claxton, K., Martin, S., Soares, M., et al. 2015. Methods for the estimation of the National Institute for Health and Care Excellence cost-effectiveness threshold. *Health Technol. Assess.* 19, 1–503.

Clayton, J.A. and Collins, F. S. 2014. Policy: NIH to balance sex in cell and animal studies. *Nature* 509, 282–3.

Clements, F. E. 1932. *Primitive Concepts of Disease*. Berkeley: University of California Press.

Clermont, G., Auffray, C., Moreau, Y., et al. 2009. Bridging the gap between systems biology and medicine. *Genome Med.* 1/9, 88.

Cohen-Haguenauer, O. 1995. Overview of gene therapy in Europe: a current statement including references to US regulation. *Hum. Gene Ther.* 6, 773–85.

Cohn, E. G., Henderson, G. E. and Appelbaum, P. S. 2017. Distributive justice, diversity, and inclusion in precision medicine: what will success look like? *Genet. Med.* 19, 157–9.

Coiera, E. 2018. The fate of medicine in the time of AI. *Lancet* 392, 2331–2.

Collins, F. 1997. Preparing health professionals for the genetic revolution. *JAMA* 278, 1285–6.

Collins, F. S. 2010. *The Language of Life: DNA and the Revolution in Personalized Medicine*. New York: HarperCollins.

Collins, F. S. 2014. Francis Collins says medicine in the future will be tailored to your genes. *Wall Street Journal*, July 7. Available at: http://www.wsj.com/articles/francis-collins-says-medicine-in-the-future-will-be-tailored-to-your-genes-1404763139

Collins, F. S. and Varmus, H. 2015. A new initiative on precision medicine. *N. Engl. J. Med.* 372, 793–5.

Conrad, P. 2007. The Medicalization of Society: On the Transformation of Human Conditions into Treatable Disorders. Baltimore: Johns Hopkins University Press.

Cook, P. A. and Bellis, M. A. 2001. Knowing the risk: relationships between risk behaviour and health knowledge. *Public Health* 115, 54–61.

Cookson, R., McCabe, C. and Tsuchiya, A. 2008. Public healthcare resource allocation and the Rule of Rescue. *J. Med. Ethics* 34, 540–4.

Coombs, C. C., Tavakkoli, M. and Tallman, M. S. 2015. Acute promyelocytic leukemia: where did we start, where are we now, and the future. *Blood Cancer J.* 5, e304.

Corral-Acero, J. et al. The 'digital twin' to enable the vision of precision cardiology. *Eur Heart J* 41, 4556–64.

Cossu, G., Birchall, M., Brown, T. et al. 2018. Lancet Commission: Stem cells and regenerative medicine. *Lancet* 391, 883–910.

Council for International Organizations of Medical Sciences. 2016. Women as Research Participants. In: *International Ethical Guidelines for Health-related Research Involving Humans*, 4th edn. Geneva: CIOMS.

Couzin-Frankel, J. 2013. Cancer immunotherapy. *Science* 342, 1432–3.

Craig, D. W. 2016. Understanding the links between privacy and public data sharing. *Nature Meth.* 13, 211.

Critchley, C., Nicol, D., Bruce, G., et al. 2019. Predicting public attitudes towards gene editing of germlines: the impact of moral and hereditary concern in human and animal applications. *Front. Genet.* 9, Article 704, doi: 10.3389/fgene.2018.00704.

Crowell, A. L., Garlow, S. J., Riva-Posse, P. and Mayberg, H. S. 2015. Characterizing the therapeutic response to deep brain stimulation for treatment-resistant depression: a single center long-term perspective. *Front. Integr. Neurosci.* 9, 41.

Cyranoski, D. 2019a. What's next for CRISPR babies. *Nature* 566, 440–2.

Cyranoski, D. 2019b. Russian scientist edits human eggs in an effort to alter deafness gene. *Nature* 575, 465–66.

Daily, L. 2019. Before you pay extra to join a concierge medical practice, consider these questions. *Washington Post*, October 2. Available at: https://www.washingtonpost.com/lifestyle/home/before-you-pay-extra-to-join-a-concierge-medical-practice-consider-these-questions/2019/10/21/90d8206a-ef8b-11e9-b648-76bcf86eb67e_story.html

Daniels, N. 2007. *Just Health: Meeting Health Needs Fairly.* Cambridge: Cambridge University Press.

David, L. A., Materna, A. C., Friedman, J., et al. 2014. Host lifestyle affects human microbiota on daily timescales. *Genome Biol.* 15, R89.

Dickenson, D. 2013. *Me Medicine vs. We Medicine: Reclaiming Biotechnology for The Common Good.* New York: Columbia University Press.

DiMai, J. A., Grabowski, H. G., Hansen, R. W. 2016. Innovation in the pharmaceutical industry: new estimates of R&D costs. *J. Health Econ.* 47, 20–33.

Djulbegovic, B., Hozo, I. and Greenland, S. 2011. Uncertainty in clinical medicine. In: Gifford, F. (ed), Philosophy of Medicine, vol. 16, pp. 299–356. Amsterdam: Elsevier B.V.

Dolgin, E. 2010. Big pharma moves from "blockbusters" to "niche busters". *Nature Med.* 16, 837.

Doudna, J. 2019. CRISPR's unwanted anniversary. *Science* 366, 777.

Downing, R. 2011. Biohealth. Beyond Medicalization: Imposing Health. Eugene, OR: Pickwick.

Drack, M., Apfalter, W. and Pouvreau, D. 2007. On the making of a system theory of life: Paul A Weiss and Ludwig von Bertalanffy's conceptual connection. *Q. Rev. Biol.* 82, 349–73.

Drake, S. 1957. *Discoveries and Opinions of Galileo*. New York: Doubleday.

Drake, S. 1966. The Accademia dei Lincei. *Science* 151, 1194–1200.

Drinkwater, C., Wildman, J. and Moffatt, S. 2019. Social prescribing. *BMJ* 364, l1285.

Dryzek, J., Nicol, D., Niemeyer, S., et al. 2020. Global citizen deliberation on genome editing. *Science* 369, 1435–7.

Dubiel, H. 2009. *Deep in the Brain: Living with Parkinson's Disease*. New York: Europa.

Duke, P., Frankel, R. M. and Reis, S. 2013. How to integrate the electronic health record and patient-centered communication into the medical visit: a skills-based approach. *Teach. Learn. Med.* 25, 358–65.

Durstewitz, D., Koppe, G. and Meyer Lindenberg, A. 2019. Deep neural networks in psychiatry. *Mol. Psychiat.* 24, 1583–98.

Duster, T. 1995. Review Essay in Symposium on the Bell Curve. *Contemporary Sociology* 24, 158–61.

Dworkin, R. 1977. *Taking Rights Seriously*. Cambridge, MA: Harvard University Press.

Dzau, V. J. and Ginsburg, G. S. 2016. Realizing the full potential of precision medicine in health and health care. *JAMA* 316, 1659–60.

Earls, J. C., Rappaport, N., Heath, L., et al. 2019. Multi-omic biological age estimation and its correlation with wellness and disease phenotypes: a longitudinal study of 3,558 individuals. *J. Gerontol. A Biol. Sci. Med. Sci.* 74, S52–S60.

Eccles, D. M., Mitchell, G., Monteiro, A. N. A., et al. 2015. BRCA1 and BRCA2 genetic testing—pitfalls and recommendations for managing variants of uncertain clinical significance. *Ann. Oncol.* 26, 2057–65.

Eckstein, L. and Nicol, D. 2020. Gene editing clinical trials could slip through Australian regulatory cracks. *J. Law Med.* 27, 274–83.

Egelie, K. J., Graff, G. D., Strand, S. P. and Johansen, B. 2016. 'The emerging patent landscape of CRISPR–Cas gene editing technology. *Nature Biotechnol.* 3, 1025–31.

El-Hazmi, M. A. 2003. Ethics of genetic counseling—basic concepts and relevance to Islamic communities. *Ann. Saudi Med.* 24, 84–92.

El Shanti, H., Chouchane, L., Badii, R., et al. 2015. Genetic testing and genomic analysis: a debate on ethical, social and legal issues in the Arab world with a focus on Qatar. *J. Transl. Med.* 13, 358.

Elligren, H. and Parsch, J. 2007. The evolution of sex-biased genes and sex-biased gene expression. *Nature Rev. Genet.* 8, 689–98.

Ellul, J. 1981. Perspectives on Our Age. Toronto: CBC.

Emanuel, E. J. 2020. Which country has the world's best healthcare? Hachette Book Group, Inc.

Emanuel E. J. and Neveloff Dubler, N. 1995. Preserving the physician–patient relationship in the era of managed care. *JAMA* 273, 323–9.

Engel, G. L. 1977. The need for a new medical model: a challenge for biomedicine. *Science* 196, 129–36.

Eriksen, J. L., Wszolek, Z. and Petrucelli, L. 2005. Molecular pathogenesis of Parkinson's disease. *Archs Neurol.* 62, 353–7.

Erme, M. 2018. *The Personality Brokers: The Strange History of Myers-Briggs and the Birth of Personality Testing.* New York: Doubleday.

Eskin, C. R. 2002. Hippocrates, kairos, and writing in the sciences. In: Sitiora, F. and Baumlin, J. S. (eds), Rhetoric and Kairos, pp. 97–113. Albany, NY: State University of New York Press.

Ethical Clearance. 2020 [cited 10 January 2020]. Available at: http://nbcpakistan.org.pk/ethical-clearance.html

European Science Foundation. 2012. *Personalised Medicine for the European Citizen— Towards More Precise Medicine for the Diagnosis, Treatment and Prevention of Disease.* Strasbourg: ESF.

Evangelista, L., Steinhubl, S. R. and Topol, E. J. 2019. Digital health care for older adults. *Lancet* 393, 1493.

Evans J. P., Meslin, E. M., Marteau, T. M. and Caufield, T. 2011. Deflating the genomic bubble. *Science* 331, 861–2.

Evans-Pritchard, E. E. 1937. Witchcraft, Oracles and Magic Among the Azande. Oxford: Clarendon Press.

Everett, M. 2004. Can you keep a (genetic) secret? The genetic privacy movement. *J. Genet. Counsel.* 13, 273–91.

Ewing, A. T., Erby, L. A., Bollinger, J., et al. 2015. Demographic differences in willingness to provide broad and narrow consent for biobank research. *Biopreserv. Biobank* 13, 98–106.

Fábrega, H. 1997. *Evolution of Sickness and Healing.* Berkeley: University of California Press.

Farley, P. 2016. $185m Gift Launches UCSF Weill Institute for Neurosciences. *UCSF News*, April 25; available at: https://www.ucsf.edu/news/2016/04/402471/185m-gift-launches-ucsf-weill-institute-neurosciences

Faltus, T. 2020. The regulation of human germline genome modification in Germany. In: Boggio, A., Romano, C. and Almqvist, J. (eds), *Human Germline Modification and the Right to Science: A Comparative Study of National Laws and Policies*, pp. 241–265. Cambridge: Cambridge University Press.

Fausto-Sterling, A. 1989. Life in the XY Corral. *Women's Stud. Int. Forum* 12, 319–31.

Feeney, O., Cockbain J., Morrison M., et al. 2018. Patenting foundational technologies: lessons from CRISPR and other core biotechnologies. *Am. J. Bioeth.* 18, 36–48.

Feero, W. G. 2017. Introducing "genomics and precision health". *JAMA* 317, 1842–3.

Feinstein, M. 1985, *Igrot Moshe, Hosen Mishpat*, p. 319, New York. [Hebrew]

Fernald, G. H., Capriotti, E., Daneshjou, R., et al. 2011. Bioinformatics challenges for personalized medicine. *Bioinformatics* 27, 1741–1748.

Ferretti, G., Linkeviciute, A. and Boniolo, G. 2017. Comprehending and communicating statistics in breast cancer screening. Ethical implications and potential solutions. In: Gadebusch-Bondio, M., Spöring, F. and Gordon, J.-S. (eds), Medical Ethics, Prediction and Prognosis: Interdipliplinary Perspectives, pp. 30–41. New York: Routledge.

Finocchiaro, M. A. 2019. *On Trial for Reason: Scienec, Religion, and Culture in the Galileo Affair*. Oxford: Oxford University Press.

Fins, J. J. 2006. Affirming the right to care, preserving the right to die: disorders of consciousness and neuroethics after Schiavo. *Support. Palliat. Care* 42, 169–78.

Fins, J. J. 2007. Border zones of consciousness: another immigration debate? *Am. J. Bioeth.* 7, 51–4.

Fins, J. 2011. Neuroethics and the lure of technology. In: Illes, J. and Sahakian, B. (eds), *Handbook of Neuroethics*, pp. 895–908. New York: Oxford University Press.

Fins, J. J. 2015. *Rights Come to Mind: Brain Injury, Ethics, and the Struggle for Consciousness*. Cambridge: Cambridge University Press.

Fins, J. J. 2016. A review of *Making Medical Knowledge* by Miriam Solomon. New York, Oxford University Press. *Notre Dame Philosophical Reviews*. March 7. Available at: https://ndpr.nd.edu/

Fins, J. J. 2016. Giving voice to consciousness. *Cambr. Q. Healthcare Ethics* 25, 583–99.

Fins, J. J. 2017. Brain injury and the Civil Right we don't think about. *New York Times*, August 24; available at: https://www.nytimes.com/2017/08/24/opinion/minimally-conscious-brain-civil-rights.html?_r=0

Fins, J. J. 2019. Disorders of consciousness in clinical practice: Ethical, legal and policy considerations. In: Posner, J. P., Saper, C. B., Schiff, N. D. and Claassen, J. (eds), *Plum and Posner's Diagnosis and Treatment of Stupor and Coma*, 5th edn, pp. 449–77. New York: Oxford University Press.

Fins, J. J. 2020. The Jeremiah Metzger Lecture: Disorders of consciousness and the normative uncertainty of an emerging nosology. *Trans. Am. Clin. Climatol. Assoc.* 131, 235–69.

Fins, J. J. and Bernat, J. L. 2018. Ethical, palliative, and policy considerations in disorders of consciousness. *Archs Phys. Med. Rehab.* 99, 1927–31.

Fins, J. J. and Plum, F. 2004. Neurological diagnosis is more than a state of mind: Diagnostic clarity and impaired consciousness. *Archs Neurol.* 61, 1354–5.

Fins, J. J. and Pohl, B. R. 2015. Neuro-palliative care and disorders of consciousness. In: Hanks, G., Cherny, N. I., Christakis, N. A., et al. (eds), *Oxford Textbook of Palliative Medicine*, 5th edn, pp. 285–91. Oxford: Oxford University Press.

Fins, J. and Reardon, J. 2019. Hannah Arendt in St. Peter's Square. *Bioethic Forum,* October 14; available at: https://www.thehastingscenter.org/hannah-arendt-in-st-peters-square/

Fins, J. J. and Schiff, N. D. 2010. In the blink of the mind's eye. *Hastings Center Rep.* 40, 21–23.

Fins, J. J. and Shapiro, Z. E. 2013. Deep brain stimulation, brain maps and personalized medicine: Lessons from the human genome project. *Brain Topogr.* 27, 55–62.

Fins, J. J., Wright, M. S. and Bagenstos, S. R. 2020. Disorders of consciousness and disability law. *Mayo Clin. Proc.* 95, 1732–9.

Fiore, R. N. and Goodman, K. W. 2015. Precision medicine ethics: selected issues and developments in next generation sequencing, clinical oncology, and ethics. *Curr. Opin. Oncol.* 28, 83–7.

Firth W. J. 1991. Chaos—predicting the unpredictable. *BMJ* 303, 1565–8.

Flexner, A. 1910. Medical Education in the United States and Canada: A Report to the Carnegie Foundation for the Advancement of Teaching. Bulletin No. 4. New York City: Carnegie Foundation for the Advancement of Teaching.

Flores, M., Glusman, G., Brogaard, K., et al. 2013. P4 medicine: how systems medicine will transform the healthcare sector and society. *Personal. Med.* 10, 565–76.

Floridi, L. 2013. The Philosophy of Information. Oxford: Oxford University Press.

FDA (Food and Drug Administration, US). 2020. *KEYTRUDA: Highlights of prescribing information.* Available at: https://www.accessdata.fda.gov/drugsatfda_docs/label/2018/125514s034lbl.pdf

FDA (Food and Drug Administration, US). 2020b. *New Drug Therapy Approvals 2019.* Available at: https://www.fda.gov/drugs/new-drugs-fda-cders-new-molecular-entities-and-new-therapeutic-biological-products/new-drug-therapy-approvals-2019

Foresman, G. A. (ed.) 2013. Aristotle's metaphysics of monsters and why we love supernatural. In: Foresman, G. A. and Irwin, W. (eds), *Supernatural and Philosophy: Metaphysics and Monsters*, pp. 72–93. Hoboken, NJ: Wiley–Blackwell.

Fortun, M. 2015. What *Toll* pursuit: affective assemblages in genomics and postgenomics. In: Richardson, S. S. and Stevens, H. (eds), *Postgenomics: Perspectives on Biology After The Genome*, pp. 32–55. Durham, NC: Duke University Press.

Francis, L. P. 2017. The significance of injustice for bioethics. *Teaching Ethics* 17, 1–8.

Frank, A. W. 1995. *The Wounded Storyteller: Body, Illness and Ethics.* Chicago: University of Chicago Press.

Freedman, B. 1987. Equipoise and the ethics of clinical research. *N. Engl. J. Med.* 317, 41–145.

Friedman, J. M., Lyons Jones, K. and Carey, J. C. 2020. Exome sequencing and clinical diagnosis. *JAMA* 324, 627–8.

Friend, T. 2021. Watch and learn. *The New Yorker.* October 25.

Fugelli, P. 2001. James Mackenzie Lecture. Trust—in general practice. *Br. J. Gen. Pract.* 51, 575–9.

Funkenstein, A. 1986. *Theology and The Scientific Imagination: From the Middle Ages to The Seventeenth Century.* Princeton: Princeton University Press.

Gadamer, H-G. 1996. The Enigma of Health. Cambridge: Polity Press.

Gaj, T., Gersbach C. A. and Barbas III, C. F. 2013. ZFN, TALEN, and CRISPR/Cas-based methods for genome engineering'. *Trends Biotechnol.* 31, 397–405.

Galen, C. 1964. Claudii Galeni Opera Omnia, editionem curavit (ed. Kühn, C. G.), 20 vols, reprinted. Hildesheim: G. Olms.

Gallacher, K., May, C. R., Montori, V. M. and Mair, F. S. 2011. Understanding patients' experience of treatment burden in chronic heart failure using normalization process theory. *Ann. Fam. Med.* 9, 235–43.

Gao, C. 2018. The future of CRISPR technologies in agriculture. *Nature Rev. Mol. Cell Biol.* 19, 275–6.

Gao, Z., Waggoner, D., Stephens, M., et al. 2015. An estimate of the average number of recessive lethal mutations carried by humans. *Genetics* 199, 1243–54.

Garber, A. M. and Tunis, S. R. 2009. Does comparative-effectiveness research threaten personalized medicine? *N. Engl. J. Med.* 360, 1925–7.

Garcia-Falgueras, A., Ligtenberg, L., Kruijver, F. P. M. and Swaab, D. F. 2011. Galanin neurons in the intermediate nucleus (InM) of the human hypothalamus in relation to sex, age and gender identity. *J. Comp. Neurol.* 519, 3061–84.

Gargon, E., Gorst, S. L., Harman, N. L., et al. 2018. Choosing important health outcomes for comparative effectiveness research: 4th annual update to a systematic review of core outcome sets for research. *PloS One* 13, e0209869.

Gaskell, G., Bard, I., Allansdottir, A., et al. 2017. Public views on gene editing and its uses. *Nature Biotechnol.* 35, 1021–3.

Geertzm C. 1973. *The Interpretations of Cultures.* New York: Basic Books.

Gefenas, E., Cekanauskaite, A., Tuzaite, E., et al. 2011. Does the "new philosophy" in predictive, preventive and personalised medicine require new ethics? *EPMA J.* 2, 141–7.

Geiss, G. K., Bumgarner, R. E., Birditt, B., et al. 2008. Direct multiplexed measurement of gene expression with color-coded probe pairs. *Nat. Biotechnol.* 26, 317–25.

Genetic Alliance UK. 2019. *Fixing the Present Building for the Future: Newborn Screening for Rare Conditions* (Draft). London: Genetic Alliance UK.

Ghaemi, S. N. 2009. The rise and fall of the biopsychosocial model. *Brit J Psychiat.* 195, 3–4.

Ghaly M. (ed) 2016. Genomics in the Gulf region and Islamic ethics. Doha, Qatar: World Innovative Summit for Health (WISH). file:///C:/Users/ymbar/Downloads/023E.pdf

Ghaly, M. 2015. Biomedical scientists as co-muftis: their contribution to contemporary Islamic bioethics. Die Welt des Islams 55, 286–311.

Ghaly, M. 2018. Islamic ethics and genomics: mapping the collective deliberations of Muslim religious scholars and biomedical scientists. In Ghaly, M. (ed.), *Islamic Ethics and the Genome Question*, pp. 47–79. Leiden: Brill.

Giacino, J. T., Ashwal, S., Childs, N., et al. 2002. The minimally conscious state: definition and diagnostic criteria. *Neurology* 58, 349–53.

Giacino, J. T., Fins, J. J., Laureys, S. and Schiff, N. D. 2014. Disorders of consciousness after acquired brain injury: the state of the science. *Nature Rev. Neurol.* 10, 99–114.

Giacino, J. T., Kalmar, K. and Whyte, J. 2004. The JFK Coma Recovery Scale—Revised: measurement characteristics and diagnostic utility. *Archs Phys. Med. Rehab.* 85, 2020–9.

Giacino, J. T., Katz, D. I., Schiff, N. D., et al. 2018. Practice guideline update recommendations summary: disorders of consciousness. *Neurology* 91, 450–60. [Simultaneously published in *Archs Phys. Med. Rehab.* 99, 1710–19.]

Gianfrancesco, M. A., Tamang, S., Yazdany, J. and Schmajuk, G. 2018. Potential biases in machine learning algorithms using electronic health record data. *JAMA Intern. Med.* 178, 1544–7.

Gibbs, W. W. 2014. Medicine gets up close and personal. *Nature* 506, 144–5.

Gibson, W. M. 1971. Can personalized medicine survive? *Can. Fam. Phys.* 17, 29–88.

Gigerenzer, G., Gaissmaier, W., Kurz-Milcke, E., et al. 2007. Helping doctors and patients make sense of health statistics. *Psychol. Sci. Public Interest* 8, 53–96.

Gilbert, S. F. 2018. Health fetishism among the Nacirema: a fugue on Jenny Reardon's The Postgenomic Condition: Ethics, Justice, and Knowledge after the Genome (Chicago University Press, 2017) and Isabelle Stengers' Another Science is Possible: A Manifesto for Slow Science (Polity Press, 2018). *Organisms J. Biol. Sci.* 2, 43–54.

Gilks, W. P., Abbott J. K. and Morrow, E. H. 2014. Sex differences in disease genetics: evidence, evolution and detection. *Trends Genet.* 30, 453–63.

Gillman, M. W. and Hammond, R. A. 2016. Precision treatment and precision prevention: integrating "below and above the skin". *JAMA Pediatrics* 170, 9–10.

Gillon, R. 2001. Confidentiality. In: Kuhse, H. and Singer, P. (eds), *A Companion to Bioethics*, pp. 513–519. Oxford: Blackwell.

Gingerich, O. 1982. The Galileo affair. *Scientific American* August.

Ginn, S. L., Amaya, A. K., Alexander, I. E., et al. 2018. Gene therapy clinical trials worldwide to 2017: an update. *J. Gene Med.* 35, e3015.

Ginsburg, G. S. and Huntington, F. W. 2009. Genomic and personalized medicine: foundation and applications. *Transl. Res.* 154, 277–87.

Giroux, É. 2015. Epidemiology and the bio-statistical theory of disease: a challenging perspective. *Theor. Med. Bioeth.* 36, 175–95.

Gladstone, I. 1932. The health examination idea. *Milbank Memorial Fund Q. Bull.* 10, 81–100.

Glasziou, P., Moynihan, R., Richards, T. and Godlee, F. 2013. Too much medicine; too little care. *BMJ* 347, f4247.

Glickman, S. W., McHutchison, J. G., Peterson, E. D., et al. 2009. Ethical and scientific implications of the globalization of clinical research. *N. Engl. J. Med.* 360, 816–23.

Goldman, M. 2016. Education in medicine: moving the boundaries to foster interdisciplinarity. *Front. Med.* 3, 15.

Goldman, S. M., Marek, K., Ottman, R., et al. 2019. Concordance for Parkinson's disease in twins. *Ann. Neurol.* 85, 600–5.

Golubnitschaja, O., Baban, B., Boniolo, G., et al. 2016. Medicine in the early twenty-first century: paradigm and anticipation—EPMA position paper 2016. *EPMA J.* 7, 23.

Good, B. J. 1994. *Medicine, Rationality and Experience: An Anthropological Perspective.* Cambridge: Cambridge University Press.

Gøtzche, P. 2012. Mammography screening: truth, lies, and controversy. *Lancet* 380, 218.

Greely, H. T. 2007. The uneasy ethical and legal underpinnings of large-scale genomic biobanks. *Annu. Rev. Genomics Hum. Genet.* 8, 343–64.

Greenwood, T. 1939. Pope Pius XI and his scientific interests. *Nature* 143, 274.

Grissinger, M. 2010. The five rights. *Pharmacy and Therapeutics* 35, 542.

Grob, R. 2019. Qualitative research on expanded prenatal and newborn screening: robust but marginalized. *Hastings Center Rep.* 41, S72–S81.

Gronde, T. van der, Uyl-de Groot, C. A. and Pieters, T. 2017. Addressing the challenge of high-priced prescription drugs in the era of precision medicine: a systematic review of drug life cycles, therapeutic drug markets and regulatory framework. *PLoS One* 12, 1–34.

Groopman, J. 2007. *How Doctors Think?* New York: Houghton Mifflin.

Guiccioli, A. 1888. *Quintino Sella.* Vol. 2. Rovigo: Officina Tipografica Minelliana.

Gymrek, M., McGuire, A. L., Golan, D., et al. 2013. Identifying personal genomes by surname inference. *Science* 339, 321–4.

Gyngell, C. Newson, A. J., Wilkinson, D., et al. 2019. Rapid challenges: ethics and genomic neonatal intensive care. *Pediatrics* 143 (Suppl. 1), S14–S21.

Haggerty, K. and Ericson, R. 2000. The surveillant assemblage. *Br. J. Sociol.* 51, 605–22.

Hahn, A. W. and Martin, M. G. 2015. Precision medicine: lessons learned from the SHIVA trial. *Lancet Oncol.* 6, e580–81.

Haines-Demont, A., Chahal, G., Bruen, A. J., et al. 2020. Testing suicide risk prediction algorithms using phone measurements with patients in acute mental health settings: feasibility study. *JMIR Mhealth Uhealth* 8, e15901.

Hall, C. T. 1996. Biotech Industry Battles Move to Ban Patents. *San Francisco Chronicle*, May 16; available at: https://www.sfgate.com/business/article/Biotech-Industry-Battles-Move-to-Ban-Patents-3032432.php

Hall, M. 2006. Researching medical trust in the United States. *J. Health Organ. Manag.* 20, 456–7.

Hall, M. A., Dugan E., Zheng B. and Mishra, A. K. 2001. Trust in physicians and medical institutions: what is it, can it be measured, and does it matter? *Milbank Q.* 79, 613–39.

Hamburg, M. and Cameron, E. 2019. WHO sticks to 2020 governance plan for human-genome editing. *Nature* 575, 287.

Hamdy, S. 2012. Our bodies belong to God: organ transplants, Islam, and the struggle for human dignity in Egypt. Berkeley: University of California Press.

Hammer, M. J. 2016. Precision medicine and the changing landscape of research ethics. Oncol. Nurs. Forum 43, 149–50.

Hansberry, D. R., Agarwal, N. and Baker, S. R. 2015. Health literacy and online educational resources: an opportunity to educate patients. *J. Roentgenol.* 204, 111–16.

Hansberry, D. R. Agarwal, N., Gonzales, S. F. and Baker, S. R. 2014. Are we effectively informing patients? A quantitative analysis of on-line patient education resources from the American Society of Neuroradiology. *Am. J. Neuroradiol.* 35, 1270–5.

Haraway, D. 2016. *Staying with the Trouble: Making Kin in the Chthulucene.* Durham: Duke University Press.

Harcourt, B. E. 2007. *Against Prediction: Profiling, Policing, and Punishing in the Actuarial Age.* Chicago: University of Chicago Press.

Harcourt, B. E. 2015. *Exposed: Desire and Disobedience in the Digital Age.* Cambridge, MA: Harvard University Press.

Harmanci, A. and Gerstein, M. 2016. Quantification of private information leakage from phenotype-genotype data: linking attacks. *Nature Meth.* 13, 251.

Harmon, A. 2018. 'Could Somebody Please Debunk This?': Writing About Science When Even the Scientists Are Nervous, *New York Times*, October 18; available at: https://www.nytimes.com/2018/10/18/insider/science-genetics-white-supremacy.html

Hart, H. L. A. 1977. *The Concept of Law.* Oxford: Oxford University Press.

Hartman, M., Martin, A. B., Benson, J. and Catlin, A. 2020. National health care spending in 2018: growth driven by accelerations in medicare and private insurance spending. *Health Affairs* 39, 8–17.

Hastings Center, New York. 2019. New project seeks to build diverse participation in precision medicine research. 30 October. Available at: https://www.thehastingscenter.org/news/new-project-seeks-to-build-diverse-participation-in-precision-medicine-research/

Hayden, E. C. 2020. If DNA Is Like Software, Can We Just Fix the Code? *MIT Technology Review*, February 26; available at: https://www.technologyreview.com/2020/02/26/905713/dna-is-like-software-fix-the-code-personalized-medicine/

Heckman-Stoddard, B. M. and Smith, J. J. 2014. Precision medicine clinical trials: defining new treatment strategies. *Semin. Oncol. Nurs.* 30, 109–16.

Hedgecoe, A. 2001. Schizophrenia and the narrative of enlightened geneticization. *Social Stud. Sci.* 31, 875–911.

Heller, M. A. and Eisenberg, R. S. 1998. Can patents deter innovation? The anticommons in biomedical research. *Science* 280, 698–701.

Hens, K., Nys, H., Cassiman, J. J. and Dierckx, K. 2011. Risks, benefits, solidarity: a framework for the participation of children in genetic biobank research. *J. Paediatr.* 158, 842–8.

Hewick, R. M., Hunkapiller, M. W., Hood, L. E., et al. 1981. A gas–liquid solid phase peptide and protein sequenator. *J. Biol. Chem.* 256, 7990–7.

Hey, S. P. and Kesselheim, A. S. 2016. Countering imprecision in precision medicine. *Science* 353, 448–9.

Hildebrandt, C. C. and Marron, J. M. 2018. Justice in CRISPR/Cas9 research and clinical applications. *AMA J. Ethics* 20, E826–833.

Hill, R. 2013. Human Genetic Commission publish final report. *BioNews* 7. https://www.bionews.org.uk/page_93623

Hisdai, A. 1824. *Ben HaMelech VeHanazir*, p. 161, Warsaw.

Hoefflin, R., Geißler, A.-L., Fritsch, R., et al. 2018. Personalized clinical decision making through implementation of a molecular tumor board: a German single-center experience. *JCO Precis. Oncol.* 2, 1–16.

Hofmann, B. 2018. Getting personal on overdiagnosis: on defining overdiagnosis from the perspective of the individual person. *J. Eval. Clin. Pract.* 24, 983–7.

Hood, L. 2008. A personal journey of discovery: developing technology and changing biology. Annu. Rev. Anal. Chem. (Palo Alto) 1, 1–43.

Hood, L. 2019. How technology, big data, and systems approaches are transforming medicine. *Res. Technol. Manage.* 62, 24–30.

Hood, L. 2019. Vision of 21st century personalized medicine. (Paper presented at the Personalized Medicine Conference, Pontifical Academy of Science, Vatican.)

Hood, L. and Auffray, C. 2013. Participatory medicine: a driving force for revolutionizing healthcare. *Genome Med.* 5, 110.

Hood, L. and Flores, M. 2012. A personal view on systems medicine and the emergence of proactive P4 medicine: predictive, preventive, personalized and participatory. *New Biotechnol.* 29, 613–24.

Hood, L. and Friend, S. H. 2011. Predictive, personalized, preventive, participatory (P4) cancer medicine. *Nat. Rev. Clin. Oncol.* 8, 184–7.

Hood, L. and Price, N. D. 2014. Demystifying disease, democratizing health care. *Sci. Transl. Med.* 6, 225ed5.

Hood, L., Price, N. and Huang, S. 2014. A vision of the future of healthcare. Institute for Systems Biology, Annual Report, pp. 10–16. Available at: https://www.isbscience.org/wp-content/uploads/ISBAnnualReport2014.pdf

Hood, L., Heath, J. R., Phelps, M. E. and Lin, B. 2004. Systems biology and new technologies enable predictive and preventative medicine. *Science* 306, 640–3.

Hood, L., Lovejoy, J. C. and Price, N. D. 2015. Integrating big data and actionable health coaching to optimize wellness. *BMC Med.* 13, 4.

Hood, L., Rowen, L., Galas, D. J. and Aitchison, J. D. 2008. Systems biology at the Institute for Systems Biology. *Brief. Funct. Genom. Proteom.* 7, 239–48.

Horvath, S. J., Firca, J. R., Hunkapiller, T., et al. 1987. An automated DNA synthesizer employing deoxynucleoside 3'-phosphoramidites. *Meth. Enzymol.* 154, 314–26.

Huang, S. and Hood, L. 2019. Personalized, precision, and n-of-one medicine: a clarification of terminology and concepts. *Perspect. Biol. Med.* 62, 617–39.

Human Genetic Commission. 2005. *Profiling the Newborn: A Prospective Gene Technology?*. London.

Human Genome Project. 2020. Availaable at: https://www.genome.gov/human-genome-project

Hunkapiller, M., Kent, S., Caruthers, M., et al. 1984. A microchemical facility for the analysis and synthesis of genes and proteins. *Nature* 310, 105–11.

Hunt, M. 1993. *The Story of Psychology.* New York: Doubleday.

Hunt, S. and Jha, S. 2015. Can precision medicine reduce overdiagnosis? *Acad. Radiol.* 22, 1040–1.

Hunter, D. J. 2016. Uncertainty in the era of precision medicine. *N. Engl. J. Med.* 375, 8.

Hwang, D., Lee, I. Y., Yoo, H., et al. 2009. A systems approach to prion disease. *Mol. Syst. Biol.* 5, 252.

Ideker, T., Dutkowski, J. and Hood, L. 2011. Boosting signal-to-noise in complex biology: prior knowledge is power. *Cell* 144, 860–3.

Ideker, T., Galitski, T. and Hood, L. 2001a. A new approach to decoding life: systems biology. *Annu. Rev. Genom. Hum. Genet.* 2, 343–72.

Ideker, T., Thorsson, V., Ranish, J. A., et al. 2001b. Integrated genomic and proteomic analyses of a systematically perturbed metabolic network. *Science* 292, 929–34.

Igo, S. E. 2018. *The Known Citizen: A History of Privacy in Modern America.* Cambridge, MA: Harvard University Press.

Illich, I. 1976. Medical Nemesis. The Expropriation of Health. London: Calder & Boyars.

Institute for Global Prosperity, University College London. 2017. Social prosperity for the future: a proposal for Universal Basic Services. London: UCL Institute for Global Prosperity. Available at: https://www.ucl.ac.uk/bartlett/igp/sites/bartlett/files/universal_basic_services_-_the_institute_for_global_prosperity_.pdf

Institute for Theological Encounter with Science and Technology. 1996. *Patenting of Biological Entities: Proceedings of the ITEST Workshop,* edited by Marianne Postiglione and Robert A. Brungs. St. Louis, Missouri: ITEST Faith/Science Press.

Institute of Medicine, Committee on Accelerating Rare Diseases Research and Orphan Product Development. 2010. Field, M. J. and Boat, T. F. (eds), Rare Diseases and Orphan Products: Accelerating Research and Development. Washington DC: National Academies Press.

Institute for Systems Biology. 2014. Pioneering the future—Seattle, WA; 2014. Available at: https://www.isbscience.org/wp-content/uploads/ISBAnnualReport2014.pdf (accessed 16 November 2020).

Interlandi, J. 2016. The paradox of precision medicine. Sci. Amer. 15, 24–5.

ITEST. 1996. Patenting of Biological Entities: Proceedings of the ITEST Workshop. St. Louis (MI).

Ito, K., Oki, R., Sekine, Y., et al. 2019. Screening for prostate cancer: history, evidence, controversies and future perspectives toward individualized screening. *Int. J. Urol.* 26, 956–970.

Jackson, F. L. 2008. Ethnogenetic layering (EL): an alternative to the traditional race model in human variation and health disparity studies. *Ann. Hum. Biol.* 35, 121–44.

Jafarey, A. M., Iqbal, S. P. and Hassan, M. 2012. Ethical review in Pakistan: the credibility gap. *J. Pak. Med. Assoc.* 62, 1354–7.

Jafarey, A. M. and Moazam, F. 2010. "Indigenizing" bioethics: the first center for bioethics in Pakistan. *Cambr. Q. Healthcare Ethics* 19, 353–62.

Jasanoff, S. (Ed.). 2011. *Reframing Rights: Bioconstitutionalism in the Genetic Age.* Cambridge: MIT Press.

Jasanoff, S. and Hurlbut, B. J. 2018. A global observatory for gene editing. *Nature* 555, 435–7.

Jennett, B. and Plum, F. 1972. Persistent vegetative state after brain damage. A syndrome in search of a name. *Lancet* 299, 734–7.

Jiang, Y., Ranney, M. L., Perez, B., et al. 2016. Burden of violent death on years of life lost in Rhode Island, 2006–2013. *Am. J. Prev. Med.* 51S, 251–9.

Joel, D. 2011. Male or female?: Brains are intersex. *Integr. Neurosci.* 5, 1–5.

Joel, D. and Fausto-Sterling, A. 2015. Beyond sex differences: new approaches for thinking about variation in brain structure and function. *Phil. Trans. R. Soc.* 371, 20150451.

Johnston, J. F., Lantos, J. D., Goldenberg, A., et al. 2018. Sequencing newborns: a call for nuanced use of genomic technologies. *Hastings Center Rep.* 41, S2–S51.

Joshi, S., Simkhada, P. and Prescott, G. J. 2011. Health problems of Nepalese migrants working in three Gulf countries. *BMC Int. Health Hum. Rights* 11, 3.

Joyner, M. J. 2015. 'Moonshot' medicine will let us down. New York Times, 29 Jan. Section A, Page 27.

Joyner, M. J. and Paneth, N. 2015. Seven questions for personalized medicine. *JAMA* 314, 999–1000.

Joyner, M. J. and Paneth, N. 2019. Promises, promises, and precision medicine. *J. Clin. Invest.* 129, 946–8.

Juengst, E. T. 1996. Self-Critical Federal Science? The Ethics Experiment within the U.S. Genome Project. *Social Philosophy and Policy* 13, 63–95.

Juengst, E. T., Flatt, M. A. and Settersten, R. A. Jr. 2012a. Personalized genomic medicine and the rhetoric of empowerment. *Hastings Center Rep.* 42, 34–40.

Juengst, E. T. and McGowan, M. L. 2018. Why does the shift from "personalized medicine" to "precision health" and "wellness genomics" matter? *AMA J. Ethics* 20, E881–890.

Juengst, E., McGowan, M. L., Fishman, J. R. and Settersten Jr, R. A. 2016. From personalized to precision medicine: the ethical and social implications of rhetorical reform in genomic medicine. *Hastings Center Rep.* 46, 21–33.

Juengst, E. T., Settersten, R. A., Fishman, J. R. and McGowan, M. L. 2012b. After the revolution? Ethical and social challenges in 'personalized genomic medicine'. *Personalized Med.* 9, 429–39.

Kagan, Y. M. 2008. *Sefer Chafetz Chaim.* New York: Mazal Press.

Kaiser, J. 2015. Obama gives East Room rollout to Precision Medicine Initiative. Science, doi: 10.1126/science.aaa6436 (30 January 2015).

Kalanithi, P. 2014. How long have I got left? *New York Times*: Opinion, 24 January. Available at: https://www.nytimes.com/2014/01/25/opinion/sunday/how-long-have-i-got-left.html (accessed 13 November 2020).

Kaldor, J., Eckstein, L., Nicol, D. and Stewart, C. 2020. Regulating innovative health technologies: dialectics, dialogics, and the case of faecal microbiota transplants. *Law Innov. Technol.* 12, 284–96.

Karns Alexander, J. 2008. *The Mantra of Efficiency: From Waterwheel to Social Control.* Baltimore: Johns Hopkins University Press.

Karow, J. 2001. Reading the book of life. *Scient. Am.* Feb. 12.

Kathiresan, S. and Srivastava, D. 2012. Genetics of human cardiovascular disease. *Cell* 148, 1242–57.

Kauffman, S., Hill, C., Hood, L. and Huang, S. 2014. Transforming medicine: a manifesto. Scientific American Worldview. Available at: https://www.medecine-integree.com/wp-content/uploads/2018/06/Transforming-Medicine_-A-Manifesto-_-worldVIEW_Kauffman.pdf

Kelly, M. P. and Barker, M. 2016. Why is changing health-related behaviour so difficult? Public Health 136, 109–16.

Kent, S. B., Hood, L. E., Beilan, H., et al. 1984. A novel approach to automated peptide synthesis based on new insights into solid phase chemistry. In: Isymiya, N. (ed.), Proceedings of the Japanese Petptide Symposium. Osaka: Protein Resarch Foundation.

Khoury, M. J. and Galea, S. 2016. Will precision medicine improve population health? *JAMA* 316, 1357–8.

Khoury, M. J., Gwinn, M. L., Glasgow, R. E., and Kramer, B. S. 2012. A population approach to precision medicine. *Am. J. Prev. Med.* 42, 639–45.

Kiddell-Monroe, R. 2014. Access to medicines and distributive justice: breaching Doha's ethical threshold. *Dev. World Bioethics* 14, 59–66.

Kim, T. 2018. Goldman Sachs Asks in Biotech Research Report: 'Is Curing Patients a Sustainable Business Model?' *CNBC*, April 11; available at: https://www.cnbc.com/2018/04/11/goldman-asks-is-curing-patients-a-sustainable-business-model.html

King, J. K. and Smith, M. E. 2016. Whole-genome screening of newborns? The constitutional boundaries of state newborn screening programs. *Pediatrics* 137 (Suppl 1), S8–S15.

Kingsmore, S. F., Petrikin, J. Willing, L. K. and Guest, E. 2015. Emergency medical genomes: breakthrough application of precision medicine. *Genome Med.* 7, 82.

Kirschstein, R. L. and Merritt, D. H. 1985. Women's Health. Report of the Public Health Service Task Force on Women's Health Issues. *Public Health Rep.* 100, 77–106.

Klauschen, F., Andreeff, M., Keilholz, U., et al. 2014. The combinatorial complexity of cancer precision medicine. Oncoscience 1, 7.

Kleiderman, E. 2020. The regulation of human germline genome modification in Canada. In: Boggio, A., Romano, C. and Almqvist, J. (eds), *Human Germline Modification and the Right to Science: A Comparative Study of National Laws and Policies*, pp. 83–102. Cambridge: Cambridge University Press.

Kleinman, A. 1997. Pain and resistance: the delegitimation and relegitimation of local worlds. In: *Writing at the Margin: Discourse Between Anthropology and Medicine*, ch. 7, pp. 169–197. Berkeley: University of California.

Kohane, I. S. 2015. Ten things we have to do to achieve precision medicine. Science 349, 37–8.

Kongsholm, N. C. H., Lassen, J. and Sandøe, P. 2018. "I didn't have anything to decide, I wanted to help my kids"—an interview-based study of consent procedures for sampling human biological material for genetic research in rural Pakistan. *Am. J. Bioeth. Empir. Bioeth.* 9, 113–27.

Kozubek, J. 2017. Who will pay for CRISPR? *Stat First Opinion.* Available at: https://www.statnews.com/2017/06/26/crispr-insurance-companies-pay/

Kraines, O. 1960. Brandeis' philosophy of scientific management. *Western Polit. Q.* 13, 191–201.

Kulkarni, R. and Soucie, J. M. 2011. Pediatric hemophilia: a review. Semin. Thromb. Hemost. 37, 737–44.

Lander, E., Baylis, F., Zhang, F., et al. 2019. Adopt a moratorium on heritable genome editing. *Nature* 567, 165–8.

Langreth, R. and Waldholz, M. 1999. New era of personalized medicine: targeting drugs for each unique genetic profile. Oncologist 4, 426–7.

Lanier, J. 2013. Who Owns the Future? New York: Simon & Schuster.

Landis Dauber, M. 2012. *The Sympathetic State: Disaster Relief and the Origins of the American Welfare State.* Chicago: University of Chicago Press.

Larson, E. J. 2021. *The Myth of Artificial Intelligence: Why Computers Can't Think The Way We Do?* Cambridge (MA): Harvard University Press.

Latour, B. and Woolgar, S. 1979. *Laboratory Life: The Social Construction of Scientific Facts.* Beverly Hills: Sage Publications.

Latour, B. and Weibel, P. (Ed.) 2005. *Making Things Public: Atmospheres of Democracy.* Cambridge, MA: MIT Press.

Laureys, S., Faymonville, M. E., Peigneux, P., et al. 2002. Cortical processing of noxious somatosensory stimuli in the persistent vegetative state. *NeuroImage* 17, 732–41.

Lázaro-Muñoz, G. and Lenk, C. 2019. The need for attention to the ethical, legal, and social implications of advances in psychiatric genomics. *Am. J. Med. Genet.* B180, 521–2.

Learning Healthcare Project. 2019. Available at: http://www.learninghealthcareproject.org/section/background/learning-healthcare-system (accessed 24 January 2021).

Ledford, H. 2019. Gene therapy's toughest test yet. Nature 576, 22–5.

Legato, M. J. (ed.) 2016. *Principles of Gender Specific Medicine: Gender in the Genomic Era,* 3rd edn. London: Academic Press.

Legato, M. (ed.) 2020. *The Plasticity of Sex.* London: Academic Press.

Lei, D., Pinaya, W. H. L., van Amelsvoort, T., et al. 2019. Detecting schizophrenia at the level of the individual: relative diagnostic value of whole-brain images, connectome-wide functional connectivity and graph-based metrics. *Psychol. Med.* 50, 1852–61.

Leite, T. K., Fonseca, R. M. C., de França, N. M. et al. 2011. Genomic ancestry, self-reported 'color' and quantitative measures of skin pigmentation in Brazilian admixed siblings. *PLoS One* 6, e27162.

Lemery, S., Keegan, P. and Pazdur, R. 2017. First FDA approval agnostic of cancer site—when a biomarker defines the indication. *N. Engl. J. Med.* 377, 1409–12.

Lepicard, E. 1998. Eugenics and Roman Catholicism an Encyclucal Letter in Context: *Casi connubii*, December 31, 1930, *Science in Context* 11:527-544.

Lepore, J. 2020. *If Then: How Simulmatics Corporation Invented The Future*. New York: Liveright.

Le Tourneau, C., Belin, L., Paoletti, X., et al. 2015a. Precision medicine: lessons learned from the SHIVA trial—Authors' reply. *Lancet Oncol.* 16, e581–2.

Le Tourneau, C., Delord, J.-P., Gonçalves, A., et al. 2015b. Molecularly targeted therapy based on tumour molecular profiling versus conventional therapy for advanced cancer (SHIVA): a multicentre, open-label, proof-of-concept, randomised, controlled phase 2 trial. *Lancet Oncol.* 16, 1324–34.

Le Tourneau, C. and Kurzrock, R. 2016. Targeted therapies: what have we learned from SHIVA? *Nat. Rev. Clin. Oncol.* 13, 719–20.

Lewis-Fernández, R. and Aggarwal, N. K. 2013. Culture and psychiatric diagnosis. *Adv. Psychosom. Med.* 33, 15–30.

Lexico.com. Definition of 'Precision'. Available at: https://www.lexico.com/definition/precision

Li, C. H., Prokopec, S. D., Sun R. X., et al. 2019. Sex Differences in Mutational Processes. https://doi.org./10.1101/528968

Li, C. H., Prokopec, S. D., Sun, R. X., et al. 2020. Sex differences in oncogenic mutational processes. *Nature Commun.* 11, 4330.

Liang, H., Lin, G., Wang, W., et al. 2020. Feasibility and safety of PD-1/L1 inhibitors for non-small cell lung cancer in front-line treatment: a Bayesian network meta-analysis. *Translat. Lung Cancer Res.* 9, 188–203.

Lichtenberg, F. R. 2021. The impact of pharmaceutical innovation on the longevity and hospitalization of New Zealand cancer patients, 1998–2017. *Expert Rev. Pharmacoecon. Outcomes Res.* 21, 476–7.

Linné, Ch. 1802. A General System of Nature Through the Three Grand Kingdoms of Animals, Vegetables, and Minerals; Systematically Divided into Their Several Classes, Orders, Genera, Species and Varieties, with Their Habitations, Manners, Economy, Structure and Peculiarities. Translated from Gmelin's last edition by William Turton: London.

Liu, L., Li, Y. and Tollefsbol, T. O. 2008. Gene–environment interactions and epigenetic basis of human diseases. *Curr. Iss. Mol. Biol.* 10, 25–36.

Liu, N., J. Sun, X. Wang, T. Zhang, M. Zhao, and H. Li. 2021. Low vitamin D status is associated with coronavirus disease 2019 outcomes: a systematic review and meta-analysis. *Int J Infect Dis.* 104, 58–64.

Livingstone, D. N. 2008. Adam's Ancestors: Race, Religion, and the Politics of Human Origins. Baltimore: Johns Hopkins University Press.

Loparco, G. 2012. Pius XII and hiding in Italy. In Bankier, D., Michman, D. and Nidam-Orvieto, I. (eds), *Pius XII and the Holocaust: Current State of Research*, pp. 115–125. Jerusalem: Yad Vashem.

Low, L. A., Mummery, C., Berridge, B. R., et al. 2021. Organs-on-chips: into the next decade. *Nature Rev. Drug Discov.* 20, 345–61.

Lown, B. A. and Rodriguez, D. 2012. Commentary: Lost in translation? How electronic health records structure communication, relationships, and meaning. *Acad. Med.* 87, 392–4.

Lunshof, J. E., Chadwick, R., Vorhaus, D. B. and Church, G. M. 2008. From genetic privacy to open consent. *Genet. Rev.* 9,406–11.

Lyotard, J. F. 1984. *The Postmodern Condition: A Report on Knowledge.* Manchester: University of Manchester Press.

McCarthy, R. M. and Arnold, A. P. 2011 Reframing sexual differentiation of the brain. *Nature Neurosci.* 14, 677–83.

MacDonald, A., Gokmen-Ozel, H., van Rijn, M. and Burgard, P. 2010. The reality of dietary compliance in the management of phenylketonuria. *J Inherit Matab Dis.* 33, 665–70.

Mackey, T. K. and Schoenfeld, V. J. 2016. Going "social" to access experimental and potentially life-saving treatment: an assessment of the policy and online patient advocacy environment for expanded access. *BMC Med.* 2, 17.

MacLeod, N. 2016. Response to Catherine Malabou, 'One life only: biological resistance, political resistance.' *Crit. Inquiry* 43, 191–9.

Maeder, M. M. and Gersbach, C. A. 2016. Genome-editing technologies for gene and cell therapy. *Mol. Ther.* 24, 430–46.

Mahase, E. 2019. Will genome testing of healthy babies save lives? *BMJ* 367, I6449.

Malik, R. 1971. The databank society: can we cope? *New Scient. Sci. J.* 49, 497–99.

Manderson, L. and Warren, N. 2010. The art of (re)learning to walk: trust on the rehabilitation ward. *Qualit. Health Res.* 20, 1418–32.

Manrai, A. K., Funke, B. H., Rehm, H. L., et al. 2016. Genetic misdiagnoses and the potential for health disparities. *N. Engl. J. Med.* 375, 655–65.

Marcum, J. A. 2008. An introductory philosophy of medicine: Humanizing modern medicine. Dordrecht, Springer.

Marcum, Z. A. and Gallad, W. A. 2012. Medication adherence to multi-drug regimens. *Clin. Geriat. Med.* 28, 287–300.

Markowitz, L. E., Liu, G., Hariri, S., et al. 2016. Prevalence of HPV after introduction of the vaccination program in the United States. *Pediatrics* 137, e20151969.

Margaliot, R. (ed) 2007. Sefer Hasidim. Jerusalem: Mosad Harav Kook.

Margalit, R. S., Roter, D., Dunevant, M. A., et al. 2006. Electronic medical record use and physician–patient communication: an observational study of Israeli primary care encounters. *Patient Educ. Counsel.* 61, 134–41.

Markett, M. J. 1996. Genetic diaries: an analysis of data protection in genetic data banks. *Suffolk Univ. Law Rev.* 30, 185–226.

Marmot, M. G., Altman, D. G., Cameron, D. A., et al. 2013. The benefits and harms of breast cancer screening: an independent review. *Br. J. Cancer* 108, 2205–40.

Maron, B. J., Maron, M. S. and Semsarian, C. 2012. Genetics of hypertrophic cardiomyopathy after 20 years: clinical perspectives. *J. Am. Coll. Cardiol.* 60, 705–15.

Marsden, G., Towse, A., Pearson, S. D., et al. 2017. *Gene Therapy: Understanding the Science, Assessing the Evidence, and Paying for Value, A Report from the 2016 ICER Membership Policy Summit.* Research Papers 001811, Office of Health Economics.

Marini-Bettolo, G. B. 1987. *The Activity of the Pontifical Academy of Sciences 1936–1986.* 2nd ed. *Scripta Varia 67.*

Mavani, H. 2014. Islam—God's deputy: Islam and transhumanism. In: Mercer, C. and Maher, D. (eds), Transhumanism and the Body, pp. 67–83. New York: Palgrave Macmillan.

May, C. and Finch, T. 2009. Implementing, embedding and integrating practices: an outline of normalization process theory. *Sociology* 43, 535–54.

May, C., Montori, V. M. and Mair, F. S. 2009. We need minimally invasive medicine. *BMJ* 339, b2803.

Mayberg, H. S., Lozano, A. M., Voon, V., et al. 2005. Deep brain stimulation for treatment-resistant depression. *Neuron* 45, 651–60.

Mayer-Schönberger, V. and Cukier, K. 2013. *Big Data: A Revolution That Will Transform How We Live, Work, and Think.* Boston: Houghton Mifflin Harcourt.

McCarthy, R. M. and Arnold, A. P. 2011. Reframing sexual differentiation of the brain. *Nature Neurosci.* 14, 677.

McCarthy, M. M., Auger, A. P., Bale, T. L., et al. 2009. The epigenetics of sex differences on the brain. *J. Neurosci.* 29, 12815–23.

McCaughey, T., Sanfilippo, P. G., Gooden, G. E. C., et al. 2016. A global social media survey of attitudes to human genome editing. *Cell Stem Cell* 18, 569–72.

McClintock, H. F. and Bogner, H. R. 2017. Incorporating patients' social determinants of health into hypertension and depression care: a pilot randomized controlled trial. *Commun. Mental Health J.* 53, 703–10.

M'charek, A., Toom, V. and Jong, L. 2020. The Trouble with Race in Forensic Identification. *Science, Technology and Human Values.* 45, 804–28.

McNally, R. 2011. What Is Mental Illness? Cambridge, MA: Harvard University Press.

Mechanic, D. and Schlesinger, M. 1996. The impact of managed care on patients' trust in medical care and their physicians. *JAMA* 275, 1693–7.

Medlineplus. 2020. Genetics. Available at: https://ghr.nlm.nih.gov/primer/hgp/elsi (accessed 23 November 2020).

Mehta, N. K., Abrams, L. R. and Myrskyla, M. 2020. U.S. life expectancy stalls due to cardiovascular disease, not drug deaths. *Proc. Natl Acad. Sci. USA* 117, 6998–7000.

Meneu, R. 2018. Life medicalization and the recent appearance of 'pharmaceuticalization'. *Farm Hosp.* 42, 174–9.

Menis, J., Hasan, B. and Besse, B. 2014. New clinical research strategies in thoracic oncology: clinical trial design, adaptive, basket and umbrella trials, new end-points and new evaluations of response. *Eur. Resp. Rev.* 23, 367–78.

Merzenich, M. M., Van Vleet, T. M. and Nahum, M. 2014. Brain plasticity-based therapeutics. *Front. Hum. Neurosci.* 8, 385.

Metcalfe, S., Hurworth, R., Newstead, J. and Robins, R. 2002. Needs assessment study of genetics education for general practitioners in Australia. *Genet. Med.* 4, 71–7.

Milanovic, B. 2019. *Capitalism Alone: The Future of the System that Rules The World.* Cambridge, MA: Harvard University Press.

Miller, V. M. 2012. In pursuit of scientific excellence: sex matters. *Adv. Physiol. Educ.* 36, 1427–8.

Miller, F. G. and Brody, H. 2003. A critique of clinical equipoise: therapeutic misconception in the ethics of clinical trials. *Hastings Center Rep.* 33, 19–28.

Miller, G. W. and Jones, D. P. 2014. The nature of nurture: Refining the definition of the exposome. *Toxicol Sci.* 137(1), 1–2.

Miller, A. M., Garfield, S., Woodman, R. C. 2016. Patient and provider readiness for personalized medicine. *Pers. Med. Oncol.* 5, 158–167.

Milstein, B., Homer, J., Briss, P., et al. 2011. Why behavioral and environmental interventions are needed to improve health at lower cost. *Health Aff. (Millwood)* 30, 823–32.

Miotto, R., Wang, F., Wang, S., et al. 2018. Deep learning for healthcare: review, opportunities and challenges. *Brief Bioinform.* 19, 1236–46.

Mittlestadt, B., Benzler J., Engelman L., et al. 2018. Is there a duty to participate in digital epidemiology? *Life Sci. Soc. Policy* 14, 9.

Moazam, F. 2006. *Bioethics and Organ Transplantation in a Muslim Society: A Study in Culture, Ethnography, and Religion*, pp. 72–73 and 122–123. Bloomington: Indiana University Press.

Moazam, F. 2006. Research and developing countries: hopes and hypes. *La Revue de Santé de la Méditerranée Orientale* 12, 30–6.

Moazam, F. 2019. "Doing Bioethics" in Pakistan. The Hastings Center. Available at: https://www.thehastingscenter.org/doing-bioethics-in-pakistan/ (accessed 16 February 2019).

Moazam, F., Jafarey, A. M. and Shirazi, B. 2014. To donate a kidney: public perspectives from Pakistan. *Bioethics.* 28, 76–83.

Moazam, F., Zaman, R. M. and Jafarey, A. M. 2009. Conversations with kidney vendors in Pakistan. *Hastings Center* 39, 29–44.

Mondal, S. and Abrol, D. 2015. Clinical trials industry in India: a systematic review. New Delhi: Institute for Studies in Industrial Development. Available at: http://isid.org.in/pdf/WP179.pdf

Morgan, C. P. and Bale, T. L. 2011. Early prenatal stress epigenetically programs dysmasculinization in second-generation offspring via the paternal lineage. *Neuroscience* 31, 11748–55.

Morin, S. H. X., Cobbold, J. F. L., Lim, A. K., et al. 2009. Incidental findings in healthy control research subjects using whole-body MRI. *Eur. J. Radiol.* 72, 529–33.

Mukherjee, S. 2017. *The Gene: An Intimate History*. New York: Scribner.

Multi-Society Task Force on PVS. 1994. Medical aspects of the persistent vegetative state (2). *N. Engl. J. Med.* 330, 1572–9.

Nahum, M., Lee, H. and Merzenich, M. M. 2013. Principles of neuroplasticity-based rehabilitation. *Prog. Brain Res.* 207, 141–71.

National Academy of Medicine, National Academy of Sciences, and the Royal Society. 2020. Heritable Human Genome Editing. Washington DC: National Academies Press.

National Academies of Sciences, Engineering and Medicine. 2017. *Human Genome Editing: Science, Ethics, and Governance*. Washington DC: National Academies Press.

National Academy of Sciences. 2011. *Toward Precision Medicine: Building a Knowledge Network for Biomedical Research and a New Taxonomy of Disease*. Washington DC: NAS.

National Academy of Sciences. 2012. *Best Care at Lower Cost: The Path to Continuously Learning Health Care in America*. Washington DC: NAS.

National Cancer Institute. 2020. Cancer Stat Facts: Non-Hodgkin Lymphoma. Available at: https://seer.cancer.gov/statfacts/html/nhl.html

National Commission for the Protection of Human Subjects of Biomedical and Behavioral Research, Department of Health, Education and Welfare (30 September 1978). The Belmont Report. Washington DC: United States Government Printing Office.

National Genome Institute (US). 2020. *DNA Sequencing Costs: Data*. Available at: https://www.genome.gov/about-genomics/fact-sheets/DNA-Sequencing-Costs-Data

National Health and Medical Research Council. 2020a. *Expert Working Committee Statement to the NHMRC CEO on the Science of Mitochondrial Donation*. Available at: https://www.nhmrc.gov.au/health-advice/all-topics/mitochondrial-donation#download

National Health and Medical Research Council. 2020b. *Report on the NHMRC's Public Consultation on the Social and Ethical Issues Raised by Mitochondrial Donation*. Available at: https://www.nhmrc.gov.au/health-advice/all-topics/mitochondrial-donation#download=

National Institutes of Health. 2020. *National Institutes of Health—All of Us*. Available at: https://allofus.nih.gov/ (accessed 23 November 2020).

National Institutes of Health, Office of Research on Women's Health. History of Women's Participation in Clinical Research. Available at: https://orwh.od.nih.gov/toolkit/recruitment/history (accessed 20 July 2020).

National Institutes of Health Office of Strategic Coordination—The Common Fund. 2019. Genotype–Tissue Expression: Expanding Our View of the Genomic Landscape Using the Genotype–Tissue Expression (GTEx) Data Set. Available at: https://commonfund.nih.gov/GTEx/ (accessed 16 February 2019).

National Institutes of Health Office of Strategic Coordination—The Common Fund. 2019. Sex as biological variable: studying sex differences in biomedical research. Available at: https://commonfund.nih.gov/sexdifferences#GTEx_Sex%20Differen ces (accessed 16 February 2019).

National Library of Medicine, US National Institutes of Health. 2020. *Online Mendelian Inheritance in Man (OMIM)*. Available at: www.ncbi.nlm.nih.gov/omim

National Research Council. 2011. Report of the Committee on Defining and Advancing the Conceptual Basis of Biological Sciences in the 21st Century, pp. 90–109. Washington DC: National Academies Press.

National Research Council. 2011. *Toward Precision Medicine: Building a Knowledge Network for Biomedical Research and a New Taxonomy of Disease*. Washington DC: National Academies Press.

National Research Council. Committee on a Framework for Developing a New Taxonomy of Disease. 2012. Toward Precision Medicine: Building a Knowledge Network Biomedical Research and a New Taxonomy of Diseases. Washington DC: National Academies Press.

Natri, H. M., Sayres, M. W. and Buetow, K. 2018. Sex-specific regulatory mechanisms underlying hepatocellular carcinoma. *bioRχiv*. https://www.biorxiv.org/content/10.1101/507939v1.full

Nebert, D. W. and Zhang, G. 2012. Personalized medicine: temper expectations. Science 337, 910.

Nelson, A., Herron, D., Rees, G. and Nachev, P. 2019. Predicting scheduled hospital attendance with artificial intelligence. *NPJ Digit. Med.* 2, 1–7.

Ng, P. C., Zhao, Q., Levy, S., et al. 2008. Individual genomes instead of race for personalized medicine. *Clin. Pharmacol. Ther.* 84, 306–9.

Ngandu, T., Lehtisalo, J., Solomon, A., et al. 2015. A 2 year multidomain intervention of diet, exercise, cognitive training, and vascular risk monitoring versus control to prevent cognitive decline in at-risk elderly people (FINGER): a randomised controlled trial. Lancet 385, 2255–63.

Nicol, D. 2020. The regulation of human germline genome modification in Australia. In: Boggio, A., Romano, C. and Almqvist, J. (eds), *Human Germline Modification and the Right to Science: A Comparative Study of National Laws and Policies*, pp. 543–67. Cambridge: Cambridge University Press.

Nicol, D., Bubela, T., Chalmers, D., et al. 2016. Precision medicine: drowning in a regulatory soup? J. Law Biosci. 3, 281–303.

Nicol, D., Eckstein, L., Morrison, M., et al. 2017. Key Challenges in bringing CRISPR-mediated somatic cell therapy into the clinic. *Genome Med.* 9, 85–8.

Niedzwiecki, M. M., Walker, D. I., Vermeulen, R., et al. 2018. The exposome: molecules to populations. *Annu. Rev. Pharmacol. Toxicol.* 59, 107–27.

Nilsson, E. Larsen, G., Manikkam, M., et al. 2012. Environmentally induced epigenetic transgenerational inheritance of ovarian disease. *PLoS One* 7, 1–18.

Noble, S. 2018. *Algorithms of Oppression.* New York: New York University Press.

Norris, J. 2020. Clues to Brain Development and Disease Emerge from 3d Epigenome Study. *UCSF News,* October 14; available at: https://www.ucsf.edu/news/2020/10/418776/clues-brain-development-and-disease-emerge-3d-epigenome-study

Nuffield Council on Bioethics. 2010. *Medical Profiling and Online Medicine: the Ethics of 'Personalised Healthcare' in a Consumer Age.* London: Nuffield Council on Bioethics.

Nuzzo, R. 2014. Scientific method: statistical errors. P values, the 'gold standard' of statistical validity, are not as reliable as many scientists assume. Nature 506, 150–152.

Nyholm, S. and O'Neill, E. 2016. Deep brain stimulation, continuity over time, and the true self. *Cambr. Q. Healthcare Ethics* 25, 647–58.

Obama, B. 2015. Precision Medicine Initiative 2015 State of the Union address. Available at: https://obamawhitehouse.archives.gov/precision-medicine (accessed 30 April 2017).

Ober, C., Loisel, D. A. and Gilad, Y. 2008. Sex-specific genetic architecture of human disease. *Nature Rev. Genet.* 9, 911–22.

Obermeyer, Z. and Emanuel, E. J. 2016. Predicting the future—big data, machine learning, and clinical medicine. *N. Engl. J. Med.* 375, 1216.

Offit, P. A. 2020. *Overkill: When Modern Medicine Goes Too Far.* New York: Harper.

O'Keefe, M., Perrault, S., Halpern J., et al. 2015. "Editing" genes: a case study about how language matters in bioethics. *Am. J. Bioeth.* 15, 3–10.

O'Leary, D. 2006. *Roman Catholicism and Modern Science.* New York: Continuum.

Ollendorf, D. A., Chapman, R. H. and Pearson, S. D. 2018. Evaluating and valuing drugs for rare conditions: no easy answers. *Value Health* 21, 547.

Omenn, G., Deangelis, C., Demets, D., et al. 2012. Evolution of translational omics: lessons learned and the path forward. Institute of Medicine Report.

Organisation for Economic Co-operation and Development. 2007. *Recommendation of the Council on the Licensing of Genetic Inventions*, OECD/LEGAL/0342. Available at: https://legalinstruments.oecd.org/en/instruments/OECD-LEGAL-0342.

Organisation for Economic Co-operation and Development. 2020. Health spending (indicator).

Ormond, K. E. and Cho, M. K. 2014. Translating personalized medicine using new genetic technologies in clinical practice: the ethical issues. *Per. Med.* 11, 211–222.

Osler, W. O. 1904. The leaven of science. In: *Aequanimitas*. London.

Osterberg L. and Blaschke, T. 2005. Adherence to medication. *N. Engl. J. Med.* 353, 487–97.

Owen, A. M., Coleman, M. R., Boly, M., et al. 2006. Detecting awareness in the vegetative state. *Science* 313, 1402.

Padela, A. I. 2015. Muslim perspectives on the American healthcare system: the discursive framing of "Islamic" bioethical discourse. *Die Welt des Islams* 55, 413–47.

Padela, A. I. 2018. Conceptualizing the human being: insights from the genethics discourse and implications for islamic bioethics. In: Ghaly, M. (ed.), Islamic Ethics and the Genome Question, pp. 111–38. Leiden: Brill.

Paine, R. 2013. I Want You to Be: Love as a Preconditiong of Thought in the Work of Hannah Arendt. In Gary Peters and Fiona Peters (eds.), *Thoughts of Love*, 107–123. Cambridge: Cambridge University Press.

Panofsky, A. and Donovan, J. 2019. Genetic ancestry testing among white nationalists: From identity repair to citizen science. *Social Studies of Science* 49, 653–81.

Papanicolas, I., Woskie, L. R. and Jha, A. K. 2018. Health care spending in the United States and other high-income countries. *JAMA* 319, 1024–39.

Parikh, R. B., Teeple, S. and Navathe, A. S. 2019. Addressing bias in artificial intelligence in health care. *JAMA* 322, 2377–8.

Pathak, Y. (ed.) 2018. *Genomics-Driven Healthcare: Trends in Disease Prevention and Treatment*. Berlin: Springer.

Pathirana, T., Clark, J. and Moynihan, R. 2017. Mapping the drivers of overdiagnosis to potential solutions. *BMJ* 358, j3879.

Pearce, C., Trumble, S., Arnold, M., et al. 2008. Computers in the new consultation: within the first minute. *Fam. Pract.* 25, 202–8.

Pellegrino, E. D. and Thomasma, D. C. 1981. *A Philosophical Basis of Medical Practice: Towards a Philosophy and Ethic of the Healing Profession*. New York: Oxford University Press.

PerMed. 2015. Shaping Europe's Vision for Personalised Medicine Strategic Research and Innovation Agenda (SRIA). Available at: http://www.permed2020.eu

Petrone, J. 2018. Danish personalized medicine initiative supported by New National Genome Center. Genome Web. Available at: https://www.genomeweb.com/info rmatics/danish-personalized-medicine-initiative-supported-new-national-gen ome-center#.XGOlblxKhPY (accessed 24 November 2021).

Petryna, A. 2007. Clinical trials offshored: on private sector science and public health. *BioSocieties* 2, 21–40.

Picchi, A. 2016. The cost of Biogen's new drug: $750,000 per patient. *CBS News*, Dec. 29.

Pinel, Ph. 1798. Nosographie philosophique, ou la méthode de l'analyse appliquée à la médecine, 2 vols. Paris: Crapelet.

Pius XI. 1936. *In Multis Solaciis*. Vatican.

Pius XII. 1959. *Discorsi ai Medici*, 6th edn. Roma: Orizzonte Medico.

Plomin, R. 2018. *Blueprints: How DNA Makes Us Who We Are*. Cambridge, MA: MIT Press.

Polack, F. P., Thomas S. J., Kitchin N., et al. 2020. Safety and efficacy of the BNT162b2 mRNA COVID-19 vaccine. *N. Engl. J. Med.* 383, 2603–15.

Polsky, A. J. 1990. *The Rise of the Therapeutic State*. Princeton: Princeton University Press.

Porphyry. 1853. Introduction (or Isagoge) to the logical Categories of Aristotle. In: Owen, O. F. (ed.), The Organon, or logical treatises of Aristotle, with the introduction of Porphyry, vol. 2, pp. 609–33. London: Henry G. Bohn.

Posada de la Paz, M., Taruscio, D. and Groft, S. C. (eds). 2017. Rare Diseases Epidemiology: Update and Overview, 2nd edn. New York: Springer.

Posner, J. B., Saper, C. B., Schiff, N. D. and Plum, F. 2007. *Plum and Posner's Diagnosis of Stupor and Coma*, 4th edn. Oxford/New York: Oxford University Press.

Posner, J. B., Saper, C. B., Schiff, N. D. and Claassen, J. 2019. *Plum and Posner's Diagnosis and Treatment of Stupor and Coma*, 5th edn. Oxford/New York: Oxford University Press.

Pot, M., Spahl, W. and Prainsack, B. 2019. The gender of biomedical data: challenges for personalised and precision medicine. *Somatechnics* 9, 170–87.

Pound, P., Britten, N., Morgan, M., et al. 2005. Resisting mecdicines: a synthesis of qualitative studies of medicine taking. *Social Sci. Med.* 61, 133–55.

Prabhu, A. V., Crihalmeanu T., Hansberry, D. R., et al. 2017. Online palliative care and oncology patient education resources through Google: do they meet national health literacy recommendations? *Pract. Radiat. Oncol.* 7, 306–10.

Prainsack, B. 2015. Is personalised medicine different? (Reinscription: The Sequel. A response to Troy Duster. *Br. J. Sociol.* 66, 28–35.

Prainsack, B. 2017. *Personalized Medicine: Empowered Patients in the 21st Century?* New York City: New York University Press.

Prainsack, B. 2018. Personalised and precision medicine: what kind of society does it take? In: Meloni, M., Cromby, J., Fitzgerald, D. and Lloyd, S. (eds), *The Palgrave Handbook of Biology and Society*, pp. 683–701. London: Palgrave Macmillan.

Prainsack, B. 2021. The meaning and enactment of openness in personalised and precision medicine. *Sci. Public Policy* 47, 647.

Precision Medicine Initative. Available at: https://ghr.nlm.nih.gov/primer/precisionmedicine/initiative.

Price, N. D., Magis, A. T., Earls, J. C., et al. 2017. A wellness study of 108 individuals using personal, dense, dynamic data clouds. *Nat. Biotechnol.* 35, 747–56.

Price, N. D., Trent, J., El-Naggar, A. K., et al. 2007. Highly accurate two-gene classifier for differentiating gastrointestinal stromal tumors and leiomyosarcomas. *Proc. Natl Acad. Sci. USA* 104, 3414–19.

Pritchard, D. E., Moeckel, F., Villa, M. S., et al. 2017. Strategies for integrating personalized medicine into healthcare practice. *Personal. Med.* 14, 141–52.

Pruss, A., Kay, D., Fewtrell, L., et al. 2002. Estimating the burden of disease from water, sanitation, and hygiene at a global level. *Environ. Health Perspect.* 110, 537–42.

Pulte, D., Jansen, L. and Brenner, H. 2020. Changes in long term survival after diagnosis with common hematologic malignancies in the early 21st century. *Blood Cancer J.* 10, 56.

Qoronfleh, M. and Dewik, N. 2017. Genomics medicine innovations: trends shaping the future of healthcare and beyond. Int. J. Adv. Res. 5, 1310–25.

Queirolo, P., Boutros, A., Tanda, E., et al. 2019. Immune-checkpoint inhibitors for the treatment of metastatic melanoma: a model of cancer immunotherapy. *Semin. Cancer Biol.* 59, 290–7.

Raghavan, S. and Vassy, J. L. 2014. Do physicians think genomic medicine will be useful for patient care? *Per Med.* 11, 424–433.

Ramaswami, R., Bayer, R. and Galea, S. 2018. Precision medicine from a public health perspective. *Annu. Rev. Public Health* 39, 153–68.

Ramos, E., Callier, S. L. and Rotimi, C. N. 2012. Why personalized medicine will fail if we stay the course. *Personalized Med.* 9, 839–47.

Rankin, S. H., Stallings, K. D. and London, F. 2005. Patient Education in Health and Illness. Philadelphia: Lippincott Williams & Wilkins.

Rappuoli, R., Pizza, M., Del Giudice, G., et al. 2014. Vaccines, new opportunities for a new society. *Proc. Natl Acad. Sci. USA* 111, 12288–93.

Reardon, J. 2005. *Race to the Finish: Identity and Governance in an Age of Genomics*, Edited by P. Rabinow. Princeton, NJ: Princeton University Press.

Reardon, J. 2007. Democratic Mis-Haps: The Problem of Democratization in a Time of Biopolitics. *Biosocieties* 2, 239–56.

Reardon, J. 2017. *The Postgenomic Condition: Ethics, Justice, Knowledge after the Genome*. Chicago: The University of Chicago Press.

Reardon, J. 2019. Ends Everlasting. *The British Journal of the History of Science* 4, 83–91.

Reardon, S. 2019. CRISPR twins might have shortened lives. *Nature* 570, 16–17.

Reardon, J., Metcalf, J., Kenney, M., et al. 2015. Science and Justice: The Trouble and the Promise. *Catalyst* 1(1). Available at: https://catalystjournal.org/index.php/catalyst/article/view/reardon_metcalf_kenney_barad

Rebok, G. W., Ball, K., Guey, L. T., et al. 2014. Ten-year effects of the advanced cognitive training for independent and vital elderly cognitive training trial on cognition and everyday functioning in older adults. *J. Am. Geriatr. Soc.* 62, 16–24.

Reinhardt, U. E. 2002. "Boutique medicine" in the US. Doctors are more interested in having high incomes than providing better health care. *BMJ* 324, 1335.

Reynolds, J. M. 2020. Health for whom? Bioethics and the challenge of justice for genomic medicine. *Hastings Center Rep.* May/June, S2–S5.

Richardson, Sarah S. 2013. *Sex Itself. The Search for Male and Female in the Human Genome*. Chicago, London, The University of Chicago Press.

Robinson, G. C. 1939. The Patient as a Person: A Study of the Social Aspects of Illness. New York and London: The Commonwealth Fund; H. Milford, Oxford University Press.

Rodriguez, L. L., Brooks, L. D., Greenberg, J. H., et al. 2013. Research ethics: the complexities of genomic identifiability. *Science* 339, 275–6.

Rose, K. M., Korzekwa, K., Brossard, D., et al. 2017. Engaging the public at a science festival: findings from a panel on human gene editing. *Sci. Commun.* 39, 250–77.

Rosen, G. 1944. *The Specialization of Medicine: With Particular Reference to Ophthalmology*. New York: Froben Press.

Rovira, J., Valera, A., Colomo, L., et al. 2014. Prognosis of patients with diffuse large B cell lymphoma not reaching complete response or relapsing after frontline chemotherapy or immunochemotherapy. *Ann. Hematol.* 94, 803–12.

Ree, A., Russnes, H. G., Heinrich D., et al. 2017. Implementing precision cancer medicine in the public health services of Norway: the diagnostic infrastructure and a cost estimate. *ESMO Open* 2, e000158.

Rehman, A., Awais, M. and Baloch, N. U. 2016. Precision medicine and low-to middle-income countries. *JAMA Oncol.* 2, 293–4.

Rehman, A., Baloch, N. U. and Janahi, I. A. 2016. Lumacaftor–ivacaftor in patients with cystic fibrosis homozygous for Phe508del CFTR. *N. Engl J. Med.* 373, 1783.

Reich, D. 2018. *Who we are and how we got here: Ancient DNA amd the new science of the human past*. Oxford/New York, Oxford University Press.

Reiser, S. 1978. Medicine and the Reign of Technology. Cambridge: Cambridge University Press.

Riccardi, A. 2003. The restorations of Pius XI and John Paul II. *The Pontifical Academy of Sciences Acta 17*, 106–114.

Richardson, S. S. 2013 *Sex Itself: The Search for Male & Female in the Human Genome*. Chicago: University of Chicago Press.

Richardson, S. S. 2012. Sexing the X: how the X became the "female chromosome". *J. Women Cult. Soc.* 37, 909–33.

Richardson, S. S. 2017. Plasticity and programming: feminism and the epigenetic imaginary. *J. Women Cult. Soc.* 43, 30–51.

Roach, J. C., Glusman, G., Smit, A. F., et al. 2010. Analysis of genetic inheritance in a family quartet by whole-genome sequencing. *Science* 328, 636–9.

Robinson, J. 2015. *In The Family Way: Illegitimacy between The Great War and The Swinging Sixties*. London: Viking.

Rodrik, D. 2011. *The Globalization Paradox: Democracy and The Future of World Economy*. New York: Norton.

Rodriguez, L. L., Brooks, L. D., Greenberg, J. H., et al. 2013. Research ethics: the complexities of genomic identifiability. *Science* 339, 275–6.

Rommen, H. A. 1947. *The Natural Law: A Study in Legal and Social History of Philosophy*. London: B. Herder Book Co.

Rose, G. 1981. Strategy of prevention: lessons from cardiovascular disease. *BMJ* 282, 1847–51.

Rose, K. M., Korzekwa, K., Brossard, D., et al. 2017. Engaging the public at a science festival: findings from a panel on human gene editing. *Sci. Commun.* 39, 250–77.

Rosenberg, E. and Zilber-Rosenberg, I. 2018. The hologenome concept of evolution after 10 years. *Microbiome* 6, 78.

Ross, L. F., Rothstein, M. A. and Clayton, E. W. 2013. Premature guidance about whole-genome sequencing. *Personalized Med.* 10, 523–6.

Rotimi, C. N. and Jorde, L. B. 2010. Ancestry and disease in the age of genomic medicine. N. Engl. J. Med. 363, 1551–8.

Rumi, J. 1997. *The Essential Rumi*, transl. Barks, C., Moyne, J., Arberry, A. J. and Nicholson R., pp. 252, 258. Secaucus, NJ: Castle Books.

Russell, S. 2019. *Human Compatible: AI and the Problem of Control*. London: Allen Lane.

Saad, E. D., Paoletti, X., Burzykowski, T. and Buyse, M. 2017. Precision medicine needs randomized clinical trials. *Nature Rev. Clin. Oncol.* 14, 317–23.

Sabatello, M. and Appelbaum, P. S. 2017. The precision medicine nation. *Hastings Center Rep.* 47, 19–29.

Sabatello, M. and Juengst, E. 2019. Genomic essentialism: its provenance and trajectory as an anticipatory ethical concern. *Hastings Center Rep.* 49, 10–18.

Sachedina, A. 1998. Human clones: an Islamic view. In: McGee, G. (ed.), The Human Cloning Debate, pp. 240–1. Berkeley: Berkeley Hills Books.

Sackett, D. 1997. Evidence-based medicine. Semin. Perinatol. 21, 3–5.

Sadowski, J. 2019. When data is capital: datafication accumulation and extraction. *Big Data and Society* 1–12.

Sainato, M. 2020. The Americans dying because they can't afford medical care. The Guardian. Available at: https://www.theguardian.com/us-news/2020/jan/07/americans-healthcare-medical-costs

Saini, A. 2019. *Superior: The Return of Race Science*. Boston: Beacon Press.

Sankar, P. and Cho, M. K., Condit, CM, et al. 2004. Genetic research and health disparities. *JAMA* 291, 2985–9.

Sauvages, F. B. 1763. Nosologia methodica sistens morborum classes, genera et species, juxta Sydenhami mentem et Botanicorum ordinem, 5 vols. Ámsterdam: Frères De Tournes.

Savulescu, J. 2001. Harm, ethics committees and the gene therapy death. *J. Med. Ethics* 27, 148–50.

Scheufele, D. A., Xenos, M. A., Howell, E. L., et al. 2017. US attitudes on human genome editing. *Science* 357, 553–4.

Schiff, N. D. 2010. Recovery of consciousness after brain injury: a mesocircuit hypothesis. *Trends Neurosci.* 33, 1–9.

Schiff, N. D. 2015. Cognitive motor dissociation following severe brain injuries. *JAMA Neurol.* 72, 1413.

Schiff, N. D., Giacino, J. T. and Fins, J. J. 2009. Deep brain stimulation, neuroethics, and the minimally conscious state. *Archs Neurol.* 66, 697–702.

Schiff, N. D., Giacino, J. T., Kalmar, K., et al. 2007. Behavioural improvements with thalamic stimulation after severe traumatic brain injury. *Nature* 448, 600–3.

Schiff, N. D., Rodriguez-Moreno, D., Kamal, A., et al. 2005. fMRI reveals large-scale network activation in minimally conscious patients. *Neurology* 64, 514–23.

Schilsky, R. L. 2014. Implementing personalized cancer care. *Nat. Rev. Clin. Oncol.* 11, 432.

Schlaepfer, T. E. and Fins, J. J. 2010. Deep brain stimulation and the neuroethics of responsible publishing. *JAMA* 303, 775.

Schnakers, C., Chatelle, C., Majerus, S., et al. 2010. Assessment and detection of pain in noncommunicative severely brain-injured patients. *Expert Rev. Neurotherap.* 10, 1725–31.

Schnakers, C., Vanhaudenhuyse, A., Giacino, J., et al. 2009. Diagnostic accuracy of the vegetative and minimally conscious state: Clinical consensus versus standardized neurobehavioral assessment. *BMC Neurology* 9, 35.

Schork, N. J. 2015. Personalized medicine: time for one-person trials. *Nature* 520, 609–11.

Schroeder, S. A. 2007. Shattuck Lecture. We can do better—improving the health of the American people. *N. Engl. J. Med.* 357, 1221–8.

Schuhmacher, A., Gassmann, O. and Hinder, M. 2016. Changing R&D models in research-based pharmaceutical companies. *J. Translat. Med.* 14, 105.

Schussler-Fiorenza Rose, S. M., Contrepois, K., Moneghetti, K. J., et al. 2019. A longitudinal big data approach for precision health. *Nat. Med.* 25, 792–804.

Schweber, S. S. 2007. *In the Shadow of the Bomb: Oppenheimer, Bethe, and the Moral Responsibility of the Scientist*. Princeton, NJ: Princeton University Press.

Science and Justice Research Center Collaborations Group. 2013. Experiments in Collaboration: Interdisciplinary Graduate Education in Science and Justice. *PLoS Biology* 11(7), 1–5.

Scutti, S. 2018. Gene Therapy for Rare Retinal Disorder to Cost $425,000 Per Eye. *CNN Health*, January 3; available at: https://www.cnn.com/2018/01/03/health/luxturna-price-blindness-drug-bn/index.html

Selby, J. V., Forsythe, L. and Sox, H. C. 2015. Stakeholder-driven comparative effectiveness research: an update from PCORI. *JAMA* 314, 2235–6.

Shachak, A. and Reis, S. 2009. The impact of electronic medical records on patient-doctor communication during consultation: a narrative literature review. *J. Eval. Clin. Pract.* 15, 641–9.

Shader, R. I. 2018. Adherence measurements in clinical trials and care. *Clin. Therapeut.* 40:1–4.

Shameer, K., Johnson, K. W., Glicksberg, B. S., et al. 2018. Machine learning in cardiovascular medicine: are we there yet? *Heart* 104, 1156–64.

Shapin, S. 1996. *The Scientific Revolution*. Chicago: University of Chicago Press.

Shapiro, S. 2007. The "Hart–Dworkin" debate: a short guide for the perplexed. In: Ripstein, A. (ed.), *Ronald Dworkin*, pp. 25–55. Cambridge: Cambridge University Press.

Sheikh, Z. A. and Hoeyer, K. 2018. "Stop talking to people; talk with them": a qualitative study of information needs and experiences among genetic research participants in Pakistan and Denmark. *J. Empir. Res. Hum. Res. Ethics* 14, 3–14.

Shendure, J., Findlay, G. M. and Snyder, M. W. 2019. Genomic medicine—progress, pitfalls, and promise. *Cell* 177, 45–57.

Sherkow, J. S. 2017. Patent protection for CRISPR: an ELSI review. *J. Law Biosci.* 4, 565–76.

Sherkow, J. S. 2019. Controlling CRISPR through law: legal regimes as precautionary principles. *CRISPR J.* 2, 299–303.

Shorter, E. 2015. The history of biopsychosocial approach in medicine: before and after Engel. In White P. (ed.), *Biopsychosocial Medicine: an integrated approach to understanding Illness*. Oxford University Press.

Shrebnivas, S. 2000. Who killed Jesse Gelsinger? Ethical issues in human gene therapy. *Monash Bioeth. Rev.* 19, 35–43.

Siddiqui, Z. S. 2003. Lifelong learning in medical education: from CME to CPD. *J. Coll. Phys. Surg. Pak.* 13, 44–7.

Siminoff, L. A., Wilson-Genderson, M., Mosavel, M., et al. 2017. Confidentiality in biobanking research: a comparison of donor and nondonor families' understanding of risks. *Genet. Test. Mol. Biomark.* 21, 171–7.

Simmel, G. 1911/1968. *The Conflict in Modern Culture and Other Essays.* New York: Teachers' College Press.

Simon, R. 2017. Critical review of umbrella, basket, and platform designs for oncology clinical trials. *Clin. Pharmacol. Ther.* 102, 934–41.

Skirbekk, H., Middelthon, A. L., Hjortdahl, P. and Finset, A. 2011. Mandates of trust in the doctor–patient relationship. *Qualit. Health Res.* 21, 1182–90.

Smarr, L. 2012. Quantifying your body: a how-to guide from a systems biology perspective. *Biotechnol. J.* 7, 980–91.

Smith, G. D. 2011. Epidemiology, epigenetics and the 'Gloomy Prospect': embracing randomness in population health research and practice. *Int. J. Epidemiol.* 40, 537–62.

Smith, L. M., Sanders, J. Z., Kaiser, R. J., et al. 1986. Fluorescence detection in automated DNA sequence analysis. *Nature* 321, 674–9.

Smith, T. C. and Thompson, T. L. 1993. The inherent, powerful therapeutic value of a good physician–patient relationship. *Psychosomatics* 34, 166.

Snyderman, R. and Yoediono, Z. 2008. Perspective: prospective health care and the role of academic medicine: lead, follow, or get out of the way. *Acad. Med.* 83, 707–14.

Solomon, M. 2015. *Making Medical Knowledge.* Oxford: Oxford University Press.

Soo-Jin Lee, S. 2021. Obligationss of the "gift": reciprocity and responsibility in precision medicine. *The American Journal of Bioethics* 21, 57–66.

Soo-Jin Lee, S., Fullerton, S. M., Saperstein, A., et al. 2019. Ethics of Inclusion: Cultivate Trust in Precision Medicine. *Science.* 364, 941–2.

Soo-Jin Lee, S., Mountain, J. and Koenig, B. 2001. The meanings of 'Race' in the new genomics: Implications for health disparities research. *Yale Journal of Health Policy, Law, and Ethics.* 1, 33–75.

Sowa, J. 2016. When does a man beget a monster? Collect. Philosoph. XIX. Available at: http://dx.doi.org/10.18778/1733-0319.19.01

Springer, K. W., Stellman, J. M. and Jordan-Young, R. M. 2012. Beyond a catalogue of differences: a theoretical frame and good practice guidelines for researching sex/gender in human health. *Soc. Sci. Med.* 74, 1817–24.

Stange, K. C. 2009. The problem of fragmentation and the need for integrative solutions. *Ann. Fam. Med.* 7, 100–3.

Stein, R. 2019. At $2.1 Million, New Gene Therapy Is the Most Expensive Drug Ever, *National Public Radio.* May 24; available at: https://www.statnews.com/2019/05/24/hold-novartis-zolgensma-approval/

Stengers, I. 2018. *Another Science Is Possible: A Manifesto for Slow Science.* London: Polity.

Stevens, D., Claborn, M. K., Gildon, B. L., et al. 2020. Onasemnogene Abeparvovec-xioi: gene therapy for spinal muscular atrophy. *Ann. Pharmacother.* 54, 1001–9.

Strange, K. 2005. The end of "naive reductionism": rise of systems biology or renaissance of physiology? *Am. J. Physiol. Cell Physiol.* 288, C968–C974.

Strathern, M. 1988. *The Gender of the Gift: Problems with Women and Problems with Society in Melanesia*. Berkley: University of California Press.

Stulp, G., Kuijper, B., Buunk, A. P., et al. 2012. Intralocus sexual conflict over human height. *Biol. Lett.* 9, 976–8.

Sudlow, C., Gallacher, J., Allen, N., et al. 2015. UK biobank: an open access resource for identifying the causes of a wide range of complex diseases of middle and old age. PLoS Med. 12, e1001779.

Sugarman, E., Nagan, N., Zhu, H., et al. 2011. Pan-ethnic carrier screening and pre-natal diagnosis for spinal muscular atrophy: clinical laboratory analysis of >72,400 specimens. *Eur. J. Hum. Genet.* 20, 27–32.

Sui-Lee, W. 2019. China Uses DNA to Track Its People, with the Help of American Exerptise. *New York Times*, February 21; available at: https://www.nytimes.com/2019/02/21/business/china-xinjiang-uighur-dna-thermo-fisher.html

Suleman, M. 2020. *Islam and Biomedical Research Ethics*. London: Routledge.

Suleman, M. 2021. Islamic perspectives on the ethics of bringing transhuman and posthuman persons into existence. In: MacKellar, C. (ed.), Ethical Creation of New Kinds of Persons. London: Bloomsbury Press.

Sulston, J. and Ferry. G. 2002. *The Common Thread: A Story of Science, Politics, Ethics, and the Human Genome*. Washington D.C.: John Henry Press.

Sun, B. B., Maranville, J. C., Peters, J. E., et al. 2018. Genomic atlas of the human plasma proteome. *Nature* 558, 73–9.

Sunder Rajan, K. 2006. *Biocapital: The Constitution of Post-Genomic Life*. Durham, NC: Duke University Press.

Sung, J., Wang, Y., Chandrasekaran, S., et al. 2012. Molecular signatures from omics data: from chaos to consensus. *Biotechnol. J.* 7, 946–57.

Szegedi M., Zelei, T., Arickx, F., et al. 2018. The European challenges of funding or-phan medicinal products. *Orphanet J. Rare Dis.* 6, 184.

Szerter, T. 2007. The Right of Registration: development, identity registration, and so-cial security—a historical perspective. *World Dev.* 35, 67–86.

Sznajder, J. I. and Ciechanover, A. 2012. Personalized medicine: the road ahead. *Am. J. Resp. Crit. Care Med.* 186, 945–7.

Szyf, M., McGowan P. and Meaney, M. J. 2008. The social environment and the epigenome. *Environ. Mol. Mutagen.* 49, 46–60.

Tabe, K. J. 2015. How can physicians prepare for precision medicine? Available at: https://personalizedmedicine.blog/2015/10/21/how-can-physicians-prepare-for-precision-medicine/ (accessed 30 April 2017).

Tal, E. 2017. Measurement in Science. In Zalta, E. N. (ed.), *The Stanford Encyclopedia of Philosophy*. Fall 2017. Metaphysics Research Lab, Stanford University. Available at: https://plato.stanford.edu/archives/fall2017/entries/measurement-science/ (accessed 20 November 2020).

Tarini, B. A., Christakis, D. A. and Welch, H. G. 2006. State newborn screening in the tandem mass spectrometry era: more tests, more false positive results. *Pediatrics* 118, 2005–26.

Tavaokkoli, S. N., Nejadsarvari, N. and Ebrahimi, A. 2015. Analysis of medical confi-dentiality from the Islamic ethics perspective. *J. Relig. Health* 54, 427–34.

Taylor, F. K. 1979. *The Concepts of Illness, Disease and Morbus*. Cambridge: Cambridge University Press.

Taylor-Sands, M. and Gyngell, C. 2018. Legality of embryonic genome editing in Australia. *J. Law Med.* 26, 356–73.

Teira, D. 2017. Testing oncological treatments in the era of personalized medicine. In: Boniolo, G. and Nathan, M., Philosophy of Molecular Medicine: Foundational Issues in Research and Practice, pp. 236–51. New York: Routledge.

Teunissen, P. W. and Dornan, T. 2008. Lifelong learning at work. *BMJ* 336, 667.

Thagard, P. 1999. *How Scientists Explain Disease*. Princeton: Princeton University Press.

Thomas, K. 2019. This new treatment could save the lives of babies, but it costs $2.1 million. *New York Times*, May 24. Available at: www.nytimes.com/2019/05/24/health/zolgensma-gene-therapy-drug.html#:~:text=But%20It%20Costs%20%242.1%20Million,-%20The%20New%20York%20Times

Thomas, K. and Abelson, R. 2019. The $6 Million Drug Claim. *New York Times*, August 25; available at https://www.nytimes.com/2019/08/25/health/drug-prices-rare-diseases.html

Tilley, C. 2007. *Democracy*. Cambridge: Cambridge University Press.

Time. 1968. Catholic Freedom V. Authority. *Time* 92(21).

Timmermans, S. and Buchbinder, M. 2010. Patients-in-waiting: living between sickness and health in the genomics era. *J. Health Soc. Behav.* 51, 408–23.

Titmuss, R. M. 1997. *The Gift Relationship: From Human Blood to Social Policy*, 3rd edn. New York: Pantheon Books.

Tomasetti, C., Lu, L. and Vogelstein, B. 2017. Stem cell divisions, somatic mutations, cancer etiology, and cancer prevention. *Science* 355, 1330–4.

Tonelli, M. R. and Shirts, B. H. 2017. Knowledge for precision medicine mechanistic reasoning and methodological pluralism. *JAMA* 318, 17.

Topol, E. 2019. *Deep Medicine: How Artificial Intelligence Can Make Healthcare Human Again*. New York: Basic Books.

Topol, E. 2019. The Topol Review: Preparing the healthcare workforce to deliver the digital future. Leeds: Health Education England.

Torkamani, A., Andersen, K. G., Steinhubl, S. R. and Topol, E. J. 2017. High-definition medicine. *Cell* 170, 828–43.

Touchot, N. and Flume, M. 2015. The payers' perspective on gene therapies. *Nature Biotechnol.* 33, 902–4.

Trading Economics. 2020. Low and middle income population total. Available at: https://tradingeconomics.com/low-and-middle-income/population-total-wb-data.html (accessed 23 November 2020).

Tremayne, S. 2009. Law, ethics and donor technologies in Shia Iran. In: Birenbaum-Carmeli, D. and Inhorn, M. C. (eds), *Assisting Reproduction, Testing Genes: Global Encounters with the New Biotechnologies*, pp. 144–163. Oxford: Berghahn Books.

Tretter, F. 2019. "Systems medicine" in the view of von Bertalanffy's "organismic biology" and system theory. *Syst Res Behav Sci.* 36, 346–62.

Tsimberidou, A. M. and Kurzrock, R. 2015. Precision medicine: lessons learned from the SHIVA trial. *Lancet Oncol.* 16, e579–e580.

Tukiainen, T., Villani, A. C., Yen, A., et al. 2017. Landscape of X chromosome inactivation across human tissues. *Nature* 550, 244–248.

Turitz Cox, D. B., Pratt, R. J. and Zhang, F. 2015. 'Therapeutic genome editing: prospects and challenges. *Nature Med.* 21, 121–131.

Turnwald, B. P., Goyer, J. P., Boles, et al. 2019. Learning one's genetic risk changes physiology independent of actual genetic risk. *Nature Hum. Behav.* 3, 48.

Tutton, R. 2012. Personalizing medicine: futures present and past. *Soc. Sci. Med.* 75, 1721–8.

Tutton, R. 2014. *Genomics and the Reimagining of Personalized Medicine.* Farnham: Ashgate Publishing.

UK Government. 2019. *The Grand Challenge Missions.* Available at: https://www.gov.uk/government/publications/industrial-strategy-the-grand-challenges/missions (accessed 20 November 2020).

UK National Screening Committee. 2015. *Review of the UK National Screening Committee (UK NSC): Recommendations.* https://assets.publishing.service.gov.uk/government/uploads/system/uploads/attachment_data/file/443953/20150602_-_Final_Recommendations.pdf

UK Supreme Court. 2017. In the matter of Charlie Gard (Permission to Appeal Hearing), Thursday 8 June 2017; Judgment of the UK Supreme Court in the Case of Charlie Gard, 19 June 2017.

UNESCO. International Declaration on Human Genetic Data 2003. Available at: http://portal.unesco.org/en/ev.php-URL_ID=17720&URL_DO=DO_TOPIC&URL_SECTION=201.html

UNESCO Cairo Office. 2011. Ethics and Law in Biomedicine and Genetics: An overview of national regulations in Arab states. UNESCO.

US Census Department. 2018. *Older People Projected to Outnumber Children for First Time in U.S. History.* Available at: https://www.census.gov/newsroon/press-releases/2018/cb18-41-population-projections.html

Valentino, L. A., Rusen, L., Elezovic, I., et al. 2014. Multicentre, randomized, open-label study of on-demand treatment with two prophylaxis regimens of recombinant coagulation factor IX in haemophilia B subjects. *Hemophilia* 20, 398–406.

Van Beers, B., Sterckx, S. and Dickenson, D. (eds) 2018. *Personalised Medicine, Individual Choice and the Common Good.* Cambridge: Cambridge University Press.

Van Diemen, C. C. Kerstjens-Frederikse WS, Bergman KA, et al. 2017. Rapid targeted genomics in critically ill newborns. *Pediatrics* 140, e20162854.

Veale, M. and Binns, R. 2017. Fairer machine learning in the real world: Mitigating discrimination without collecting sensitive data. *Big Data Soc.* 4, 2053951717743530.

Veltman, J. A. and Brunner, H. G. 2012. De novo mutations in human genetic disease. *Nature Rev. Genet.* 13, 565–75.

Venter, J. C. 2010. Multiple personal genomes await. *Nature* 464, 676–7.

Venter, J. C. Adams, M. D., Myers, E. W., et al. 2001. The sequence of the human genome. *Science* 291, 1304–51.

Vilain, E. 2003. Expert Interview Transcripts. Rediscovering biology: molecular to global perspectives. Oregon Public Broadcasting. Available at: https://www.learner.org/series/rediscovering-biology-molecular-to-global-perspectives/human-evolution/expert-interview-transcripts-christopher-wills-phd/

Vinti, C. 2003. Federico Cesi, the first academy and Umbria. *The Pontifical Academy of Sciences Acta* 17, 41–80.

Vogt, H. 2017. Systems Medicine as a Theoretical Framework for Primary Care Medicine. Dissertation, Norwegian University of Science and Technology.

Vogt, H., Green, S. Ekstrøm, C. and Brodersen, J. 2019. How precision medicine and screening with big data could increase overdiagnosis. *BMJ* 366, l5270.

Vogt, H., Hofmann, B. and Getz, L. 2016. The new holism: P4 systems medicine and the medicalization of health and life itself. *Med. Health Care Phil.* 19, 307–23.

Vogt, H., Ulvestad, E. Eriksen, T. E. and Getz, L. 2014. Getting personal: Can systems medicine integrate scientific and huanistic conceptions of the patient? *J. Eval. Clin. Pract.* 20, 942–52.

Wade, L. 2018. To Overcome Decades of Mistrust, a Workshop Aims to Train Indigenous Researchers to Be Their Own Genome Experts. *Science*, September 27; available at: https://www.sciencemag.org/news/2018/09/overcome-decades-mistrust-workshop-aims-train-indigenous-researchers-be-their-own

Wade, N. 1972. Salk Institute: Elitist Pursuit of Biology with a Conscience. *Science* 178(4063), 846–49.

Wainberg, M., Magis, A. T., Earls, J. C., et al. 2020. Multiomic blood correlates of genetic risk identify presymptomatic disease alterations. *Proc. Natl Acad. Sci. USA* 117, 21813–20.

Waldenberg, E. Y. 1985. Responsa "Tzitz Eliezer", vol. 16, 4. Jerusalem. [Hebrew]

Waldenberg, E. Y. 1988. Responsa "Tzitz Eliezer", vol. 17, 48. Jerusalem. [Hebrew]

Walkley S. U., Davidson, C. D., Jacoby, J., et al. 2016. Fostering collaborative research for rare genetic disease: the example of Niemann–Pick Type C Disease. *Orphanet J. Rare Dis.* 11, 161.

Walsh, R. et al. 2017. Reassessment of Mendelian gene pathogenicity using 7,855 cardiomyopathy cases and 60,706 reference samples. *Genetics in Medicine* 19, 192–203.

Wang, Y., Liu, H. and Sun, Z. 2017. Lamarck rises from his grave: parental environment-induced epigenetic inheritance in model organisms and humans. *Biol. Rev.* 92, 2084–111.

Warren, B. J. 2013. How culture is assessed in the DSM-5. *J. Psychosoc. Nurs. Ment. Health Serv.* 51, 40–5.

Watson, M. S. et al.; American College of Medical Genetics Newborn Screening Expert Group. 2006. Newborn screening: toward a uniform screening panel and system—executive summary. *Pediatrics* 117 (Suppl 3), S296–S307.

Weber, M. 1917. Der Sinn der "Wertfreiheit" der soziologischen und ökonomischen Wissenschaften 1917. Engl. transl.: Weber, M., The Methodology of the Social Sciences. New York: The Free Press; 1946.

Webster, J. P., Molyneux, D. H., Hotez, P. J. and Fenwick, A. 2014. The contribution of mass drug administration to global health: past, present and future. *Phil. Trans. R. Soc.* 369B, 20130434.

Wegwarth, O. I., Schwartz, L. M., Woloshin, S., et al. 2012. Do physicians understand cancer screening statistics? A national survey of primary care physicians in the United States. *Ann. Intern. Med.* 156, 340–9.

Weiner, A. B. 1992. Inalienable Possessions: The Paradox of Keeping While Giving. Berkeley: University of California Press.

Weisberg, S. M., Badgio, D. and Chatterjee, A. 2017. A CRISPR new world: attitudes in the public toward innovations in human genetic modification. *Front. Public Health* 5, 117.

Weiss, G. J. 2015. Precision medicine: lessons learned from the SHIVA trial. *Lancet Oncol.* 16, e580.

Weisz, G. 2003. The emergence of medical specialization in the nineteenth century. *Bull. Hist. Med.* 77, 536–74.

Welch, G. and Burke, W. 2015. Why whole genome testing hurts more than it helps. Los Angeles Times, 27 April. Available at: http://www.latimes.com/opinion/op-ed/la-oe-welch-problems-predictive-medicine-20150428-story.html (accessed 30 April 2017).

West, K. M., Blacksher, E. and Burke, W. 2017. Genomics, health disparities, and missed opportunities for the nation's research agenda. *JAMA* 317, 1831–2.

Westergaard, D., Moseley, P., Karuna, F. et al. 2019. Population-wide analysis of differences in disease progression patterns in men and women. https://doi.org/10.1038/s41467-019-08475-9.

Westergaard, D., Moseley, P., Sørup, F. K. H., et al. 2019. Population-wide analysis of differences in disease progression patterns in men and women. *Nature Commun.* 15, 6669.

Weston, A. D. and Hood, L. 2004. Systems biology, proteomics, and the future of health care: toward predictive, preventative, and personalized medicine. *J Proteome Res.* 3, 179–96.

White, A. and Danis, M. 2013. Enhancing patient-centered communication and collaboration by using the electronic health record in the examination room. *JAMA* 309, 2327–8.

White House. 2015. Remarks by the President on Precision Medicine, The White House, Office of the Press Secretary. Press release, 30 January 2015.

Wigderson, A. 2019. *Mathematics and Computation: A Theory Revolutionizing Technology and Science.* Princeton: Princeton University Press.

Wikipedia. 2020. List of Countries by GNI (Nominal) Per Capita. Available at: https://en.wikipedia.org/wiki/List_of_countries_by_GNI_ (nominal,_Atlas_method)_per_capita (accessed 23 November 2020).

Williams, S. A., Kivimaki, M., Langenberg, C., et al. 2019. Plasma protein patterns as comprehensive indicators of health. *Nat. Med.* 25, 1851–7.

Willig, L. K., Petrikin, J. E., Smith, L. D., et al. 2015. Whole-genome sequencing for identification of Mendelian disorders in critically ill infants: a retrospective analysis of diagnostic and clinical findings. *Lancet Resp. Med.* 3, 377–387.

Wilmanski, T., Rappaport, N., Earls, J. C., et al. 2019. Blood metabolome predicts gut microbiome alpha-diversity in humans. *Nat. Biotechnol.* 37, 1217–28.

Wilner, M. D. 1973. Responsa Hemdat Zvi, vol. 3, 43 Jerusalem.

Wilson, J. 2009. Not so special after all? Daniels and the social determinants of health. *J. Med. Ethics* 35, 3–6.

Wilson, J. 2012. On the value of the intellectual commons. In: Lever, A. (ed.), *New Frontiers in the Philosophy of Intellectual Property*, pp. 122–139. Cambridge: Cambridge University Press.

Wilson, J. 2016. The right to public health. *J. Med. Ethics* 42, 367–75.

Wilson, J. 2021. *Philosophy for Public Health and Public Policy: Beyond The Neglectful State*. Oxford: Oxford University Press.

Wilson, J. M. G. and Jungner, G. 1968. *Principles and Practice of Screening for Disease*. Geneva: WHO.

Wiseman, M. 2010. Communicating genetic risk information within families: a review. *Famil. Cancer* 9, 691–703.

Wizemann, T. M. and Pardue, M. (eds) 2001. *Exploring the Biological Contributions to Human Health: Does Sex Matter?* Washington DC: National Academic Press.

Wolbring, G. and Diep, L. 2016. The discussions around precision genetic engineering: role of and impact on disabled people. *Laws* 5, 37; doi:10.3390/laws5030037.

Wolken, D. J. 2018. "Thinking in the Gap: Hannah Arendt and the Prospects for a Postsecular Philosophy of Education." in *Philosophy of Education 2016*, edited by N. Levinson. Urbana, Illinois: Philosophy of Education Society.

Wolkenhauer, O., Kolch, W. and Cho, K. H. 2004. Mathematical systems biology: genomic cybernetics. In: Paton R., Bolouri, H., Holcombe, M., Parish, J. H. and Tateson, R. (eds), Computation in Cells and Tissues. Natural Computing Series. Berlin: Springer.

Wolpe, P. R. 2009. Personalized medicine and its ethical challenges. *World Med. Health Policy* 1, 47–55.

World Health Organization. 2019. Statement on governance and oversight of human genome editing. Available at: https://www.who.int/news-room/detail/26-07-2019-statement-on-governance-and-oversight-of-human-genome-editing

World Health Organisation. 2021a. *Human Genome Editing: Position Paper*. Available at:https://www.who.int/publications/i/item/9789240030404

World Health Organisation. 2021b. *Human Genome Editing: Recommendations*. Available at: https://www.who.int/publications/i/item/9789240030381

World Health Organisation. 2021c. *Human Genome Editing: A Framework for Governance*. Available at: https://www.who.int/publications/i/item/9789240030060

World Health Organization. International Classification of Diseases (ICD 11). Available at: https://icd.who.int/en.

World Health Organization, World Intellectual Property Organization, and World Trade Organization. 2013. *Promoting Access to Medical Technologies and Innovation: Intersections between Public Health, Intellectual Property and Trade*. Geneva: WHO.

Worldometers. 2020. *Population By Country 2020—Worldometer*. Available at: https://www.worldometers.info/world-population/population-by-country/ (accessed 23 November 2020).

World Medical Association. 2013. World Medical Association Declaration of Helsinki: ethical principles for medical research involving human subjects. *JAMA* 310, 2191–4.

Worthey, E. Mayer, A. N., Syverson, G. D., et al. 2011. Making a definitive diagnosis successful clinical application of while exome sequencing in a child with intractable inflammatory bowel disease. *Genet. Med.* 13, 255–62.

Wright, G. W., Huang, D. W., Phelan, J. D., et al. 2020. A probabilistic classification tool for genetic subtypes of diffuse large B cell lymphoma with therapeutic implications. *Cancer Cell* 37, 551–68.

Wright, M. S., Varsava, N., Ramirez, J., et al. 2018. Justice and seere brain injury: legal remedies for a marginalized population. *Florida State Univ. Law Rev.* 45, 313–82.

Wu, J. 2020. *The Limits of Screening*. Thesis, University of Cambridge. doi: 10.17863/CAM.53085.

Wu, A. H. B., WhiteSam, O, H. and Burchard, E. 2015. The Hawaii clopidogrel lawsuit: the possible effect on clinical laboratory testing. *Personalized Med.* 12, 179–81.

Wulf, C. 2008. Mimetic learning. *Designs for Learning* 1, 56–67.

Wysham, N. G., Abernethy, A. P. and Cox, C. E. 2014. Setting the vision: applied patient reported outcomes and smart, connected digital healthcare systems to improve patient-centered outcomes prediction in critical illness. *Curr. Opin. Crit. Care* 20, 566–72.

Xu, S., Jayaraman, A. and Rogers, J. A. 2019. Skin sensors are the future of health care. *Nature* 571, 319–21.

Yang, X., Schadt, E. E., Wang, S., et al. 2006. Tissue-specific expression and regulation of sexually dimorphic genes in mice. *Genome Res.* 16, 995–1004.

Ye, L., Wang, C., Hong, L. et al. 2018. Programmable DNA repair with CRISPRa/i enhanced homology-directed repair efficiency with a single Cas9. *Cell Discov.* 4, 46–58.

Yeary, K. H., Alcaraz, K. I., Ashing, K. T., et al. 2020. Considering religion and spirituality in precision medicine. *Transl. Behav. Med.* 10, 195–203.

Young, A. H. 2020. Golodirsen: first approval. *Drugs* 80, 329–33.

Zadeh, L. A. 1972. Fuzzy languages and their relation to human and machine intelligence. In: Proc. Int. Conf. Man and Computer., Bordeaux 1970, pp. 130–165. Basel: Karger.

Zahm, J. A. 1896. *Evolution and Dogma*. Chicago: McBride & Co.

Zardavas, D. and Piccart-Gebhart, M. 2015. Clinical Trials of Precision Medicine through Molecular Profiling: Focus on Breast Cancer. ASCO Educational Book.

Zhang, Y. S. 2019. The ultimate in personalized medicine: your body on a chip. IEEE Sprectrum, 21 March 2019. Available at: https://spectrum.ieee.org/biomedical/diagnostics/the-ultimate-in-personalized-medicine-your-body-on-a-chip

Zhou, W., Sailani, M. R., Contrepois, K., et al. 2019. Longitudinal multi-omics of host-microbe dynamics in prediabetes. *Nature* 569, 663–71.

Zilberstein, Y. 2000. Tuvch Yabiuo, vol. 2. Beni Brak. [Hebrew]

Zubair, N., Conomos, M. P., Hood, L., et al. 2019. Genetic predisposition impacts clinical changes in a lifestyle coaching program. *Sci. Rep.* 9, 6805.

Zubiri, X. 1980. On Essence. Washington DC: Catholic University of America Press.

Zuboff, S. 2019. *The Age of Surveillance Capitalism: The fight for a Human Future at the New Frontier of Power*. New York: Public Affairs.

Index